UNIVERSITY OF LIVER
013520562
KU-626-420

WITHDRAWN
FROM STOCK

Antenna Design With Fiber Optics

For a complete listing of the *Artech House Antenna Library*,
turn to the back of this book.

Antenna Design With Fiber Optics

A. Kumar

Artech House
Boston • London

Library of Congress Cataloging-in-Publication Data
Kumar, A. (Akhileshwar)
Antenna design with fiber optics/A. Kumar.
p. cm.
Includes bibliographical references and index.
ISBN 0-89006-759-7 (alk. paper)
1. Antennas (Electronics)—Design and construction. 2. Fiber optics. 3. Beam optics. I. Title.
TK7871.6.K7946 1996
621.382'4—dc20 95-49990
CIP

British Library Cataloguing in Publication Data
Kumar, A. (Akhileshwar)
Antenna design with fiber optics
1. Antennas (Electronics) 2. Fiber optics 3. Antennas (Electronics)—Design
I. Title
621.3'84135

ISBN 0-89006-759-7

Cover design by Christine Koch.

685 Canton Street
Norwood, MA 02062

International Standard Book Number: 0-89006-759-7
Library of Congress Catalog Card Number: 95-49990

10 9 8 7 6 5 4 3 2 1

This book is dedicated to my family,
Nirmala, Anne, and Angel

Contents

Preface

This is the first book written about the application of optical fibers in antenna beamforming/scanning networks. It provides the theory and practical design data for the benefit of engineers, managers, and scientists in the field of antenna beamforming network using optical fiber technology. This book can be understood by those with only a slight background in optical fiber and antennas.

Chapter 1 describes the background and history of the optical fiber technology since 1950s. General performance, including losses of various transmission lines are compared with the optical fiber for application in the beamforming network technology.

Chapter 2 concentrates on fiber types, fabrication, radiation, and reliability. Various materials and processes have been described for fabrication of optical fibers. Some electrical and mechanical properties of fibers are described. The effect of nuclear radiation is also presented for application of fibers in space antennas. The crack growth due to fatigue in silica glass fiber is described.

Chapter 3 presents a brief description of various technologies used for RF/microwave beamforming/scanning for phased array antennas. These technologies are: dielectric lenses, waveguide lenses, Ruze lenses, bootlace lenses, Rotman lenses, R-2R lenses, R-KR lenses, Butler and Blass matrices, and single and dual reflector imaging antennas. Electronically controlled true time delay and the injection-phase-locking concepts are discussed for the scanning of phased array antennas.

Chapter 4 summarizes the work on the acousto-optics time-delay networks for phased array antennas. Brief descriptions of the optical fiber delay line and the background of the acousto-optics are given. Experimental and theoretical results for the optical time-delay in transmit and receive mode for the optical beamforming/scanning networks are presented.

Chapter 5 describes optically controlled beam scanning for the active phased array antennas using the method of injection locking.

Chapter 6 provides hardware-compressive fiber optical delay line for steering of one- and two-dimensional phased array antennas.

Chapter 7 presents a two-laser model for the optical beam steering and sidelobe suppression of phased array antennas.

The author wishes to thank Anne Kumar (electrical engineering student of Carleton University, Ontario) and Angel Kumar for drafting the artwork. It would not have been possible to finish this book on time without Anne's help. I also thank Helmut E. Schrank for reviewing the manuscript and Mark Walsh, Kimberly Collignon, and Beverly Cutter of Artech House for their help during preparation of the book.

The author is grateful to the Society of Photo-Optical Instrumentation Engineers, the Optical Society of America, the Institute of Electrical and Electronics Engineers, the Institute of Electrical Engineers, Horizon House, AK Electromagnetique, Inc., and Artech House for their kind permission to use figures and materials from their various publications. I also express my sincere appreciation to T. Ikegami, B. Wedding, A. H. Al-Ani, A. L. Cullen, M. Kam, J. Wilcox, Michael G. Blankenship, Charles, W. Deneka, Koichi Inada, Suzanne R. Nagel, J. B. MacChesney, Kenneth L. Walker, J. T. Krause, M. M. Bubnov, M. J. Matthewson, G. A. Koepf, Daniel Dolfi, E. N. Toughlin, J. P. Huignard, M. Baril, H. Zmuda, P. Kornreich, L. Gessell, T. M. Turpin, P. Liao, R. A. York, T. Itoh, J. Lin, S. Kawasaki, T. Berceli, R. C. Compton, A. S. Daryoush, X. Zhang, X. Zhou, P. R. Herczfeld, Nabeel A. Riza, H. D. Hristov, S. E. Lipsky, C. H. Cox III, G. E. Betts, L. M. Johnson, R. A. Soref, A. M. Levine, S. K. Chew, T. K. Tong, C. Wu, Akis P. Goutzoulis, D. Keu Davies, J. M. Zomp, Willie W. Ng, D. Yap, A. Narayanan, R. Hays, Andrew A. Walston, P. Hrycak, A. Johnson, Gregory L. Tangonan, Jar Juch Lee, Irwin L. Newberg, Norman Bernstein, Y. Konishi, W. Chujo, H. Iwasaki, K. Yasukawa, M. Fujise, J. J. Pan, I. Chiba, W. Z. Li, and other authors whose work has been used in this book.

Akhileshwar Kumar
February 1996

Chapter 1
Introduction

1.1 BACKGROUND AND HISTORY OF OPTICAL FIBER TECHNOLOGY

Telecommunication systems that use light as a transmission medium are older forms of technology than wireless and radio systems. Evidence of light-guidance technology was first demonstrated in a water stream leaving a pail in the 1850s. In the 1870s Alexander Graham Bell [1] investigated optical communications in the form of what he called the "photo phone." The theoretical basis for modern optical fiber systems was laid out in 1910 during the development of the electromagnetic theory of microwave propagation on a dielectric rod [2]. During 1930 to 1950, bundles of glass filament were fabricated to demonstrate optical fiber experiments. The first development of a GaAs p-n junction diode was demonstrated in 1962 by three groups of inventors, namely, Nathan et al. [3], Quist et al. [4], and Hall et al. [5]. The p-n junction used in these lasers was made by diffusing zinc into n-type wafers of GaAs. Pellets were cut out of them with two faces carefully polished parallel to each other and perpendicular to the plane of the junction. The two parallel, semireflective end faces form a Fabry–Perot resonant cavity that enhances the optical Q of the system. The other set of faces is roughened or sawed to suppress all but the modes propagating between the end faces. These laser diodes operated at a temperature of 77K and a wavelength of 0.85 μm.

In the same year (1962), Holonyak et al. [6] reported 77-K operation of GaAsP laser diodes at shorter wavelengths, and laser emissions at other wavelengths followed. These laser diodes were homojunction broad-area devices made by diffusion of impurities to form the p-n junction in the bulk single crystals. In their development, a p-n junction converted input electrons into photons and an optical cavity with two cleaved faces of a crystal reflected back some of the emitted photons. It operated into the infrared light region with high efficiency at a sufficient high input current. These lasers were small in size (less than a millimeter in length), reliable, durable, easy for mass production, and superior to gas lasers. In 1963, Kroemer [7] proposed that semiconductor laser diodes might be improved substantially by making use of a heterostructure. A new alloy, GaAlAs, was developed

(1968) that fabricated injection laser diodes [8–10]. The highest efficiencies and lowest threshold current densities in room temperature for a laser emitting in this general spectral range were obtained by Nelson and Kressel [11] using the p-p-n heterojunction structure. The diodes were fabricated by the sequential growth of first n and then p AlGaAs layers on a (100) GaAs substrate by liquid phase epitaxy from Ga solutions. Zn was the acceptor ($2 \times 10^{19}\text{cm}^{-3}$) and Te the donor ($2-3 \times 10^{18}\text{cm}^{-3}$). The melt compositions were adjusted to produce greater Al content (and hence greater bandgap energy) in the p^+ region than in the n-layer. Following the epitaxial growth, the diodes were heat-treated to move (by Zn diffusion) the p-n junction a distance of 1μm to 3μm into the n-layer, thus forming the optical waveguide in which the radiative recombination occurs.

It was first proposed by Kao and Hockham [12] in 1966 that highly transparent fibers might be used to carry light beams over distances of 1 km or more. Many researchers were working in the United Kingdom, Germany, France, Italy, United States, and Japan to demonstrate the possibilities of optical fiber communications. Graded-index multimode fiber, which provides a loss of about 1 dB/m, was fabricated in 1968. During 1969 and 1970, several 850-nm optical fiber components like high-radiance light-emitting diodes (LEDs), continuously operating room-temperature ILDs, and silicon (Si) avalanche photodiodes (APDs) were developed. A loss of 20-dB/km single-mode optical fiber at a wavelength of 633 nm was fabricated. Optical and microwave techniques were initially combined during the developmental work of the laser from 1962 to 1970. In the 1970s, the direction of research on optical fiber communication using optical fibers was determined in a report by AT&T laboratories where CW lasing of a laser diode at room temperature was described. Since then, innovations of the transmission systems have been performed with the progress of laser diode technology. Oxide-insulated strip geometry double-heterostructure GaAs/GaAlAs lasers have been made and evaluated at Standard Telecommunications Laboratories (STL), United Kingdom, as sources for optical communication systems. In 1976, Goodwin et al. [13] described the fabrication of the devices, including the liquid phase epitaxy and device-processing stages. Attention was given to some details that were important in controlling the device performance parameters. Evaluation of the lasers has included studies of reliability, performance at elevated temperatures, spatial mode structure, special properties, and modulation performance.

In the optical fiber transmission, it is necessary to launch enough optical power into the core of the optical fiber and to reduce the loss due to the optical fiber. The laser diode has an advantage of more than two orders of magnitude over the light-emitting diode. Figure 1.1 shows the improvement of the threshold current versus years of research. Alferov [14] proposed a GaAs/Ga (As,P) high-voltage rectifier and a DH laser in 1967. He suggested that the threshold of heterojunction lasers should be considerably lower than homojunction lasers because of injected carrier confinement and light guidance, so that it might be useful at room temperature. CW operation of laser diode is made possible by using the GaAs/AlGaAs double-heterostructure (DH). Figure 1.1 shows clearly that the quantum-well (QW) structure laser diode requires low current-density operation and that the stained-layer quantum-well (SL-QW) structure laser diode reduces the current density even further.

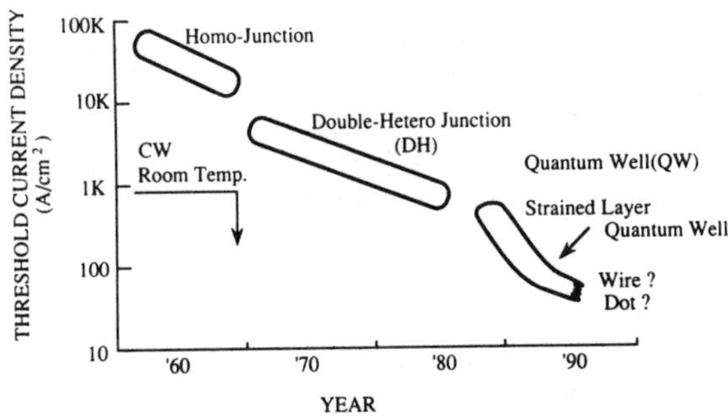

Figure 1.1 Threshold current density of laser diodes achieved versus time of research. (© 1992 IEEE.)

During the 1970s, the 1.5-Mb/s PCM copper cable system was used widely for interoffice communication due to its cost effectiveness. In time as the demand increased, digital long-distance and high-capacity systems were desired over analog ones because of their compatibility with telecommunication networks. Ikegami [15] described that as the repeater spacing is even, the most innovative 400-Mb/s coaxial cable system developed by NTT (Japan) at that time was only 1.5 km at the most. The advent of the optical fiber transmission systems was found to be the fastest. A basic system of the optical fiber is shown in Figure 1.2.

In the case of the present 400-Mb/s system, the optical power in the single-mode fiber at the sending end is 0 dBm (1 mW). The photodiode is −40 dBm (0.1 μW); the difference in the power budget, which is 40 dB, is to be dissipated evenly in the optical fiber. Assuming the loss in the fiber is 0.5 dB/km, the signal can be transmitted through the fiber for more than 80 km. With some margin taken into consideration, the signal can be sent over 50 km without a repeater. It is shown in Section 1.2 that the loss in the optical fiber is the smallest compared to other transmission media. The Appendix shows the major developments in optical fiber technology including sources and components. The enormous progress of erbium-doped optical fiber amplifier (EDFA) technology has led to a breakthrough in system developments and has opened up a new era in multigigabit optical transmission systems design as shown in Figure 1.3 [16]. This figure shows the improvement in optical transmission capacity since 1975. The devepment work on optical transmission capacity at AT&T (Bell Laboratories) has shown 2.4-Gb/s signal transmission using a circulating loop of 90 km with four *Er-doped fiber amplifiers* (EDFAs) up to 21000 km equivalently at the wavelength of 1.551 μm and the figure of merit is about 10^6. In 17 years, the improvement that has been achieved in the figure of merit is about three orders of magnitude. During this period many innovations have been performed and these include the changes in wavelength from 0.88 μm to 1.3 μm to 1.5 μm, the change of the fiber structure from multimode waveguide to single-mode waveguide, and the addition of the coherent detection format to the direct detection format.

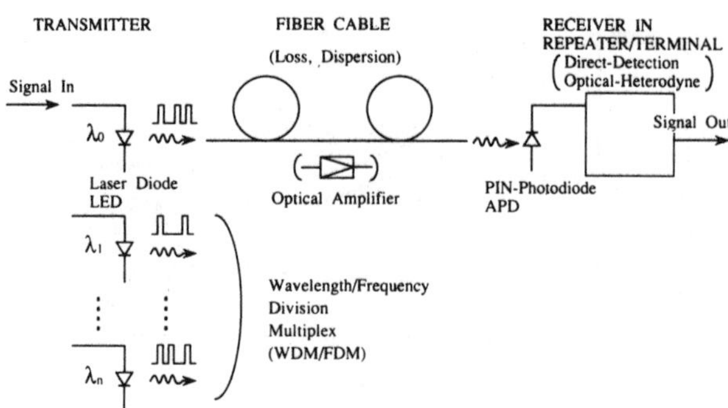

Figure 1.2 A basic optical fiber transmission system. (© 1992 IEEE.)

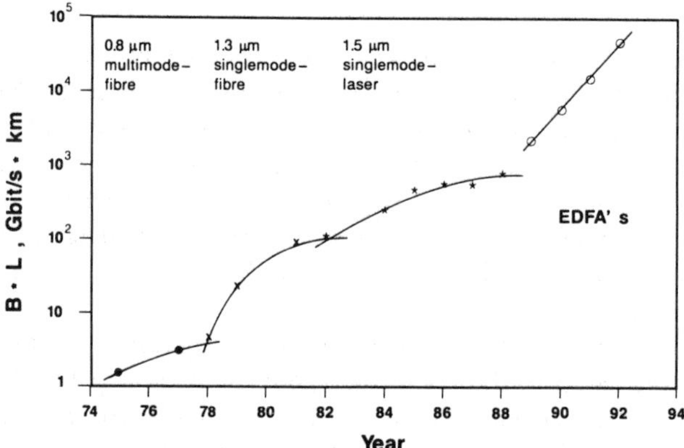

Figure 1.3 Progress of optical transmission capacity versus time. (© 1992 OSA.)

In 1987 Kumar [17] discussed that the fiber optics R&D in many research establishments indicated new developments at the components and subsystem level in the following areas: single frequency, single spatial mode, high-sensitivity photo-receivers, monolithic integration, very high speed electronics, highly stable advanced lasers, optical amplifiers, midinfrared fibers, optical repeaters, coherent optical technology, and optical sensors and detectors. Reductions in the loss of fibers that are fabricated of heavy glass are distinct possibilities in 1995. Theoretical calculations show a loss of 3×10^{-4} dB/km for this fiber versus 0.16 dB/km for silica-based fiber at 1.55 μm. It is encouraging to note the developments in midinfrared optical fibers. The prediction made by Kumar [17] was correct on the development of low-loss optical fiber undersea cables [18]. Thiennot et al. [18] described the completion of undersea optical fiber cable links in the Atlantic. Most of the work on low-loss undersea cables in the Atlantic will be completed by 1995, 1996.

Satellite communication systems have become a mature technology, along with fiber optic cables, terrestrial microwaves, and other radio media. The major area of growth in satellite usage from the 1980s to the mid-1990s is in new applications made feasible by an increasing number of K-band satellites, from extending optical fibers in the antenna feeder network and connecting private corporation networks by VSAT terminals.

It has become increasingly apparent that the next generation of electronically scanned array antenna will require smaller and higher performance signal distribution and time-delay beamforming networks. The latter characteristic provides wide instantaneous bandwidth at each steering angle, thus eliminating beam squint and enabling narrow pulse operation on large antennas, multiple-frequency operation, and multifunction aperture operation. Photonics technology has the potential to make a tremendous impact on the architecture and realization of these systems. Optical interconnects are recognized to provide wider bandwidth, lower loss, smaller size, lighter weight, and higher signal isolation than electrical transmission lines. Photonics device and circuit technology can implement modulation and true-time beamforming functions on the microwave-modulated lightwave signals and provide much wider bandwidth than is presently possible with monolithic microwave integrated circuit technology.

1.2 GENERAL PERFORMANCE CONSIDERATIONS

Transmission lines are required between the source and antenna in any system. Usually, two basic technologies (coaxial lines and waveguides) are used in the satellite antennas as transmission lines. For years satellite antennas have also used microstrip and stripline technologies. The beamforming networks of many antennas are also designed using coaxial cables, waveguides, microstrips, and stripline components. The optical fiber technology is very promising for transmission lines and beamforming network technology. In the following paragraph, we consider the following five transmission technologies:

1. Coaxial cables;
2. Waveguides;
3. Microstrip line;
4. Stripline;
5. Optical fiber.

1.2.1 Coaxial Cables

There are two types of losses in a coaxial cable: (1) conductor loss and (2) dieletric loss. The expression to calculate these losses is given in any standard electromagnetics book [19–23]. Figure 1.4 shows the geometry of a coaxial cable. The inner and outer radii of the coaxial cable are r and R, respectively. The medium between the inner and outer conductors is filled with a dielectric of relative permittivity, ϵ_r. The attenuation due to conductor loss in decibels per meter is given by

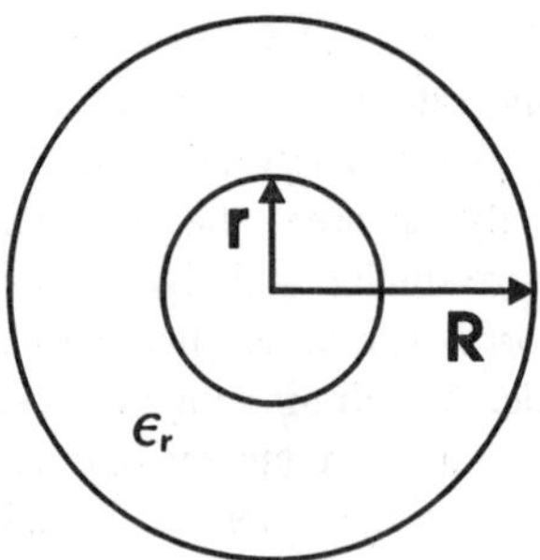

Figure 1.4 Geometry of a coaxial cable.

$$\alpha_c = \frac{8.68 R_s}{4\pi Z_0}\left(\frac{1}{R} + \frac{1}{r}\right) \text{ dB/m} \tag{1.1}$$

where, Z_0 is the characteristic impedance in Ohms and R_s is the surface resistivity,

$$Z_0 = \frac{60}{\sqrt{\epsilon_r}} \ln (R/r)\ \Omega \tag{1.2}$$

$$R_s = \sqrt{\frac{\pi f \mu_0}{\sigma}} \tag{1.3}$$

f is the frequency in Hertz, $\mu_0 = 4\,\Pi \times 10^{-7}$ H/m, σ is the conductivity of the conductor in mhos per meter (siemens), and relative permittivity is given by

$$\epsilon_\gamma = \frac{\epsilon}{\epsilon_0} = \epsilon' - i\epsilon'' \tag{1.4}$$

where ϵ' denotes the real part of (ϵ_r), ϵ'' the imaginary part of (ϵ_r), and ϵ_0 the permittivity of free space. The attenuation due to the dielectric loss is given by

$$\alpha_d = \frac{8.68 \pi f \sqrt{\epsilon_r}}{c} \tan\delta \text{ dB/m} \tag{1.5}$$

where $c = 3 \times 10^8$ m/s and tan δ is the loss tangent of the dielectric material. The loss tangent is calculated from

$$\tan\delta = \frac{\epsilon''}{\epsilon'} \tag{1.6}$$

1.2.2 Waveguides

The hollow-pipe waveguides are most commonly used as transmission lines in connecting the antennas (transmit/receive) in any satellite system or tower-mounted antennas. There are two main types of waveguides that are used: (1) rectangular and (2) circular. In satellites, manufacturers are using reduced-height waveguides to reduce the weight. These waveguides can support *transverse electric* (TE) and *transverse magnetic* (TM) waves but not TEM waves. The attenuation loss of a metallic waveguide is called metallic loss (if a waveguide is made of copper, it is called copper loss).

Rectangular Waveguide

Figure 1.5 shows the coordinate system for a rectangular waveguide. Many books [19–23] have been written on waveguides for a basic understanding. The TE_{10} mode (dominant mode) is used in a rectangular waveguide for the propagation of electromagnetic waves. The inside width and height of a rectangular waveguide is denoted by a and b, respectively. The attenuation of electromagnetic waves in a rectangular waveguide operating in TE_{mn} modes can be given by

$$\alpha_{TE_{mn}} = \frac{2R_s}{b\eta\sqrt{1-(f_c/f)^2}}\left\{\left(1+\frac{b}{a}\right)\left(\frac{f_c}{f}\right)^2 + \left[1-\left(\frac{f_c}{f}\right)^2\right]\left[\frac{(b/a)\,(b/a\,m^2+n^2)}{b^2m^2/a^2+n^2}\right]\right\} \tag{1.7}$$

The attenuation of a rectangular waveguide in the fundamental mode (TE_{10}) is given by

$$\alpha_{TE_{10}} = \frac{R_s}{b\eta\sqrt{1-(f_c/f)^2}}\left[1+\frac{2b}{a}\left(\frac{f_c}{f}\right)^2\right] \tag{1.8}$$

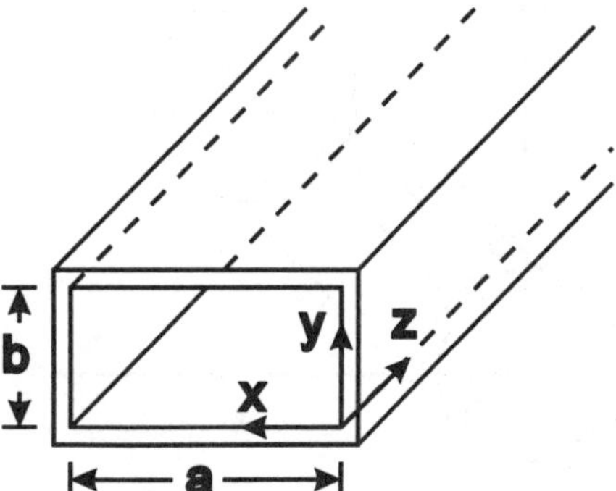

Figure 1.5 Coordinate system of a rectangular waveguide.

where R_s is defined in (1.3), m is the half-sine wave variation in the x direction ($m = 1$ for the TE_{10} mode), n is the half-sine wave variation in the y direction ($n = 0$ for the TE_{10} mode), f_c is the cut-off frequency, f is the frequency of operation $\eta = \sqrt{\mu_0}/\sqrt{\epsilon_0}$.

Circular Waveguide

Figure 1.6 shows the coordinates of a hollow-pipe (metal) circular waveguide. The inside radius of the waveguide is denoted by a, n describes the number of variations in electromagnetic waves circumferentially, and l describes the number of variations in electromagnetic waves radially in a circular waveguide. The attenuation due to transverse electric waves in the TE_{nl} mode is given by

$$\alpha_{TE_{nl}} = \frac{R_s}{a\eta} \frac{8.686}{\sqrt{1 - (f_c/f)^2}} \left[\left(\frac{f_c}{f} \right)^2 + \frac{n^2}{p'^2_{nl} - n^2} \right] \tag{1.9}$$

The attenuation of a circular waveguide in the TE_{11} mode is given by

$$\alpha_{TE_{11}} = \frac{R_s}{a\eta} \frac{1}{\sqrt{1 - (f_c/f)^2}} \left[\left(\frac{f_c}{f} \right)^2 + 0.42 \right] \tag{1.10}$$

where f is the frequency of operation in hertz, n describes the number of variations circumferentially of E (transverse magnetic wave) in a circular waveguide, l describes the number of variations radially of H (transverse electric wave) in a circular waveguide, and

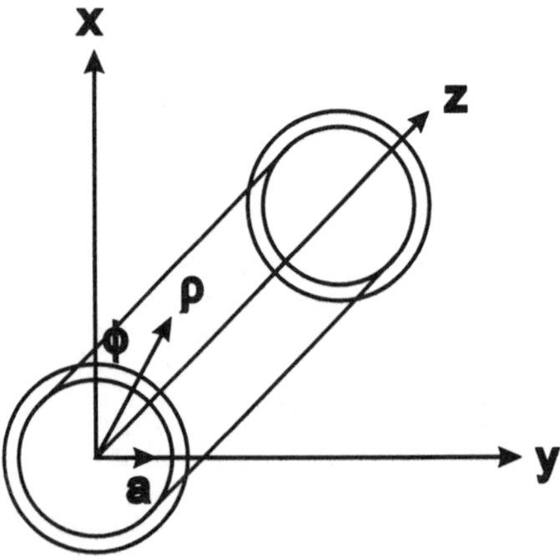

Figure 1.6 Coordinate system of a circular waveguide.

f_c is the cut-off frequency of a circular waveguide in hertz. For transverse electric waves, the required boundary condition can be satisfied if

$$J_n'(k_c a) = 0 \tag{1.11}$$

Therefore, if p_{nl}' is the lth root of $J_n'(x) = 0$ (1.11) is satisfied by

$$(k_c)_{nl} = \frac{p_{nl}'}{a} \tag{1.12a}$$

For all TE_{nl} wave type, the cut-off wavelength and frequency are given by

$$(\lambda_c)_{TE_{nl}} = \frac{1}{\sqrt{\mu_0 \epsilon_0}\,(f_c)_{TE_{nl}}} = \frac{2\pi a}{p_{nl}'} \tag{1.12b}$$

The cut-off wavelength for the TE_{11} mode in a circular waveguide is calculated by using $p_{nl}' = 1.84$ in (1.12b) as

$$(\lambda_c)_{TE_{11}} = 3.41a \tag{1.12c}$$

1.2.3 Microstrip

The microstrip line is a transmission line that connects transmit or receive antennas. Figure 1.7 shows the geometry of a microstrip transmission line. In the figure, the line width is W, the thickness of the substrate is H, the relative permittivity of the substrate is ϵ_r, and it is assumed that the thickness of the conductor on the substrate is zero (very thin). There are two main losses in the line: (1) conductor loss and (2) dielectric loss.

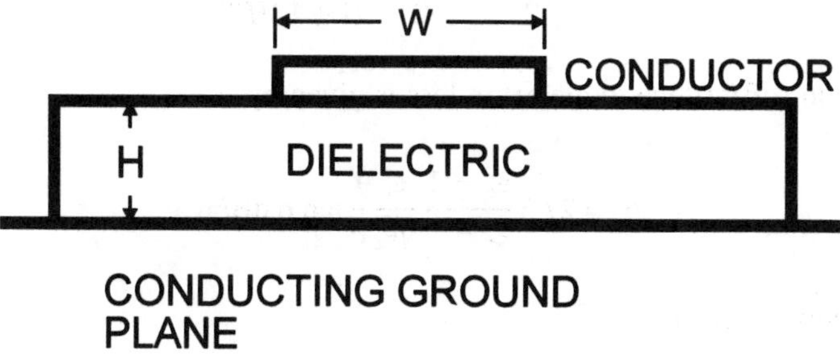

Figure 1.7 Geometry of a microstrip transmission line.

These two losses contribute to the attenuation of the line. The attenuation due to conductor loss (dB/m) is given by [22–24]

$$\alpha_c = \frac{8.68R_s}{WZ_0} \text{ dB/m} \tag{1.13}$$

where R_s is the surface resistivity and is calculated from (1.3); Z_0 is the characteristic impedance of the microstrip line and is dependent on the width, height, and the effective permittivity of the substrate. The characteristic impedance of the line is calculated from the following expressions

For $W/H < 1$

$$Z_0 = \frac{377}{2\pi\sqrt{\epsilon_e}} \ln(8H/W + W/4H)\ \Omega \tag{1.14}$$

For $W/H > 1$

$$z_0 = \frac{377}{\sqrt{\epsilon_e}}\left[\frac{1}{W/H + 1.393 + 0.667 \ln (W/H + 1.444)}\right]\Omega \tag{1.15}$$

where effective permittivity (ϵ_e) is given by the following expressions.

For $W/H < 1$

$$\epsilon_e = \frac{\epsilon_r + 1}{2} + \frac{\epsilon_r - 1}{2}\frac{1}{\sqrt{1 + 12H/W}} + 0.04(1 - W/H)^2 \tag{1.16}$$

For $W/H > 1$

$$\epsilon_e = \frac{\epsilon_r + 1}{2} + \frac{\epsilon_r - 1}{2}\frac{1}{\sqrt{1 + 12H/W}} \tag{1.17}$$

The attenuation (dB/m) due to dielectric loss is given by

$$\alpha_d = 27.3\frac{\epsilon_e - 1}{\epsilon_r - 1}\frac{\epsilon_r}{\sqrt{\epsilon_e}}\frac{f}{c} \tan\delta \text{ dB/m} \tag{1.18}$$

Total attenuation is given by

$$\alpha_t = \alpha_c + \alpha_d \text{ dB/m} \tag{1.19}$$

1.2.4 Stripline

The geometry of the stripline is shown in Figure 1.8. This type of line has been used in antennas as a transmission line in receive and transmit modes. In the figure, W is the width of the center conductor, b is the separation between the ground planes, t is the thickness of the conducting strip, and ϵ_r is the relative permittivity of the medium. The coaxial and the stripline rely on conducting ground planes separated by a dielectric material to confine a field created by exciting the center conductor. Both lines propagate in the TEM mode, and both have similar electromagnetic properties. The attenuation due to the conductor loss is given by [23]

$$\alpha_c = \frac{0.0231\, R_s \epsilon_r}{30\pi(b - t)} Z_0(A + B) \tag{1.20}$$

where

$$A = 1 + \frac{2W}{b - t} + \frac{1}{\pi}\frac{b + t}{b - t} \ln\left(\frac{2b - t}{t}\right) \tag{1.21}$$

For $Z_0 \sqrt{\epsilon_r}\,(1 + 2.3\, t/b) < 120\ \Omega$

$$B = 0 \tag{1.22}$$

and for $Z_0 \sqrt{\epsilon_r}\,(1 + 2.3\, t/b) \geq 120\ \Omega$

$$B = \frac{(0.35 - W/b)}{(b - t)\,(1 + 12\, t/b)^2} \tag{1.23}$$

$$\left[\frac{t}{b}(17.45b + 35W) - 9W + 5.85 - 32.4t^2/b\right] \tag{1.24a}$$

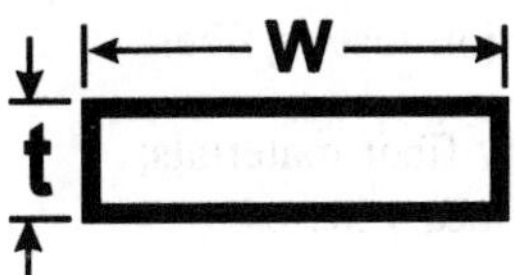

Figure 1.8 Geometry of a stripline.

The characteristic impedance is given by

$$Z_0 = \frac{30\pi}{\sqrt{\epsilon_r}} \frac{(1 - t/b)}{W_e/b + C_f/\pi} \Omega \tag{1.24b}$$

where C_f is a fringing capacitance and is calculated from

$$C_f = 2 \ln \left(\frac{1}{1 - t/b} + 1 \right) - \frac{t}{b} \ln \left\{ \frac{1}{(1 - t/b)^2} - 1 \right\} \tag{1.25}$$

W_e is the effective width of the conductor and is defined as

$$\frac{W_e}{b} = \frac{W}{b} \quad \text{if } W/(b - t) > 0.35 \tag{1.26}$$

$$\frac{W_e}{b} = \frac{W}{b} - \frac{(0.35 - W/b)^2}{1 + 12\, t/b} \quad \text{if } W/(b - t) < 0.35 \tag{1.27}$$

The attenuation due to dielectric losses is given by

$$\alpha_d = \frac{8.68\pi f \sqrt{\epsilon_r}}{c} \tan \delta \tag{1.28}$$

Total attenuation is calculated by adding the attenuations due to conductor and dielectric losses.

1.2.5 Optical Fiber

The use of optical fiber for signal transmission in conventional communications has been increasing rapidly. The transmission loss is the most important of the optical properties of the fiber. The system cost is significantly controlled by loss. There are four mechanisms that are responsible for transmission loss in fiber:

1. Losses due to absorption by fiber materials;
2. Scattering (linear and nonlinear) losses;
3. Bending losses;
4. Core and cladding losses.

Losses Due to Absorption by Fiber Materials

The following are mechanisms of absorption by fiber materials:

- Atomic defects in fiber materials;
- Extrinsic absorption by impurity atoms in fiber materials;
- Loss due to the basic constituent atom of the fiber.

Atomic Defects in Fiber Materials

Atomic defects of the atomic structure of the fiber material cause absorption loss. These atomic defects of the fiber material are the missing molecules, oxygen defects in glass structures, and high-density clusters of atoms. Losses due to these defects are negligible compared to other material absorption. These losses are significant in the earth's Van Allen Belts (VAB) due to alfa, beta, and gamma bombardments discussed by Kumar [25]. The effect of radiation on the fiber for low-orbit satellite antennas is very significant. A detailed study of the losses and life of the fiber optic materials for low-, medium-, and high-orbit satellite antennas is provided by Kumar [25, 26].

Extrinsic Absorption by Impurity Atoms in Fiber Materials

Extrinsic absorption by transition metal impurities that are present in the starting metals (iron, chromium, cobalt, copper, and from OH (water) ions) used for direct-melt fibers range between 1 part and 10 parts per billion, causing losses from 0.001 dB/m to 0.01 dB/m [27]. The main reason for the impurity absorption losses is electronic transitions between the energy levels associated with the incompletely filled inner subshell of these ions.

Loss Due to the Basic Constituent Atom of the Fiber

Intrinsic absorption is associated with physical properties in the basic fiber material. The loss results from electronic absorption bands in the ultraviolet region and from atomic vibration bands in the near-infrared region. The attenuation is given by the empirical relationship [28, 29]

$$\alpha_{uv} = C\, e^{(E/E_0)} \tag{1.29}$$

where E is the photon energy and C and E_0 are the empirical constants. An empirical formula for the infrared attenuation (dB/m) for GeO_2-SiO_2 glass fiber is given by Miya et al. [30] and Nagel et al. [31] as

$$\alpha_{IR} = 7.81 \times 10^{11} \times \exp\left(\frac{-48.48}{\lambda}\right) \quad \text{dB/m} \tag{1.30}$$

Figure 1.9 shows the optical fiber attenuation characteristics and their limiting mechanisms for a GeO_2-doped low-loss low-OH-content silica fiber [32]. Curves of attenuation in the ultraviolet and the infrared regions and scattering losses are also given in the figure. The measured loss of a single-mode fiber is plotted by solid line in the figure. A comparison of the infrared absorption induced by various doping materials in low-OH-content fibers is also shown in Figure 1.9.

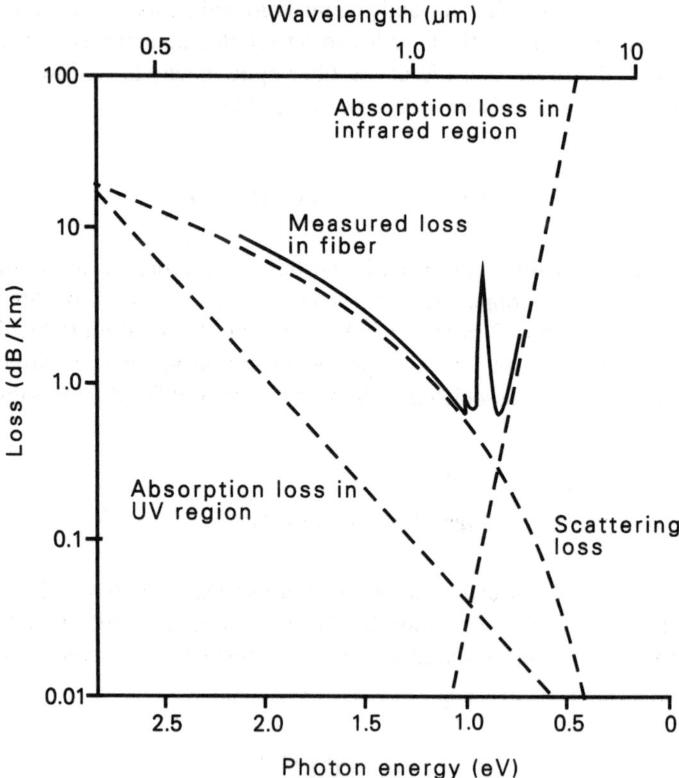

Figure 1.9 Attenuation characteristics and their limiting mechanisms for a GeO_2-doped low-loss low-OH-content silica fiber. (© 1976 IEE.)

Scattering (Linear and Nonlinear) Losses

This loss in optical fiber is produced due to compositional fluctuations, variations in density, defects during manufacturing, and structural inhomogeneities. Olshansky [33] and Maurer [34] gave the following approximate formulae to calculate the attenuation due to scattering:

$$\alpha_{scat} = \frac{8\pi^3}{3\lambda^4}(n^2 - 1)^2 k_B T_f \beta_T \tag{1.31}$$

where n is the refractive index, k_B is Boltzmann's constant, β_T is the isothermal compressibility of the fiber material, and T_f is the temperature at which the density fluctuations occur.

Bending Losses

There are two types of bends that occur in optical fibers when laying out an antenna or transmission system: (1) a bending radius that is greater than the fiber diameter and (2) random bends of the fiber axis (it happens when the fibers are incorporated into cables). Figure 1.10 shows calculated and measured bending losses at a wavelength of

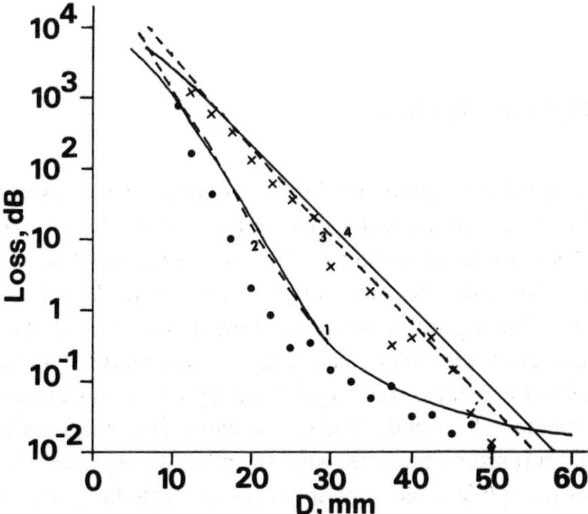

Figure 1.10 Calculated and measured bending losses for matched-clad and depressed-clad fiber at $\lambda = 1.55$ μm [36]: (1) Depressed-clad fibers (Eq. (5) of [36]); (2) Depressed-clad fibers (conventional formula of [38]); (3) Matched-clad fibers (conventional formula of [38]); (4) Matched-clad fibers (Eq. (5) of [36]): measured points for depressed-clad (from [37]) and measured points for matched-clad fibers (from [37]).

1.55 μm for matched-clad [35] and depressed-clad fibers [36]. As the loop diameter decreases, the loss increases exponentially. Another form of radiation loss is due to nonuniformities in the manufacturing of the fiber and by nonuniform lateral pressures created during the cabling of the fiber. This type of loss can be minimized by using a straightjacket over the fiber.

Core and Cladding Losses

The core and cladding provide losses because they have different indices of refraction. Gloge [39] reported the loss due to core and cladding for a step-index waveguide operating in a mode of order (*v*, *m*);

$$\alpha_{vm} = \alpha_1 \frac{P_{\text{core}}}{P} + \alpha_2 \frac{P_{\text{clad}}}{P} \tag{1.32}$$

where α_1 is the attenuation coefficient for the core, α_2 is the attenuation coefficient for the cladding, and P_{core}/P and P_{clad}/P are the fractional powers.

1.3 COMPARISON OF LOSSES

A major factor in the rapidly growing popularity of optical fiber is its very low loss, which has been reduced by more than an order of magnitude over the past decade and is now below 0.003 dB/m for systems operating at 0.85 μm (less than 0.0005 dB/m at 1.3 μm). Calculations of losses for microstrip and striplines are more than 1 dB, which is out of scale for Figure 1.11. The figure shows loss comparisons of optical fibers with some ordinary metallic guiding media: foam coax, coax in vacuum (center conductor supported by thin low-loss dielectric pieces), rectangular waveguides, and circular waveguide operating in the fundamental TE_{11} mode (5-cm diameter). The combination of low loss and small size is at first surprising because with metallic waveguides the general trend is toward lower loss as the guide cross section increases. This happens because waveguide losses are dominated by resistive dissipation near the surface of the conductors, and bigger cross sections generally lead to lower resistance. In the case of optical fiber, the loss is a bulk effect in the dielectric medium, essentially independent of the cross section, so the fiber can be made very small without incurring added loss.

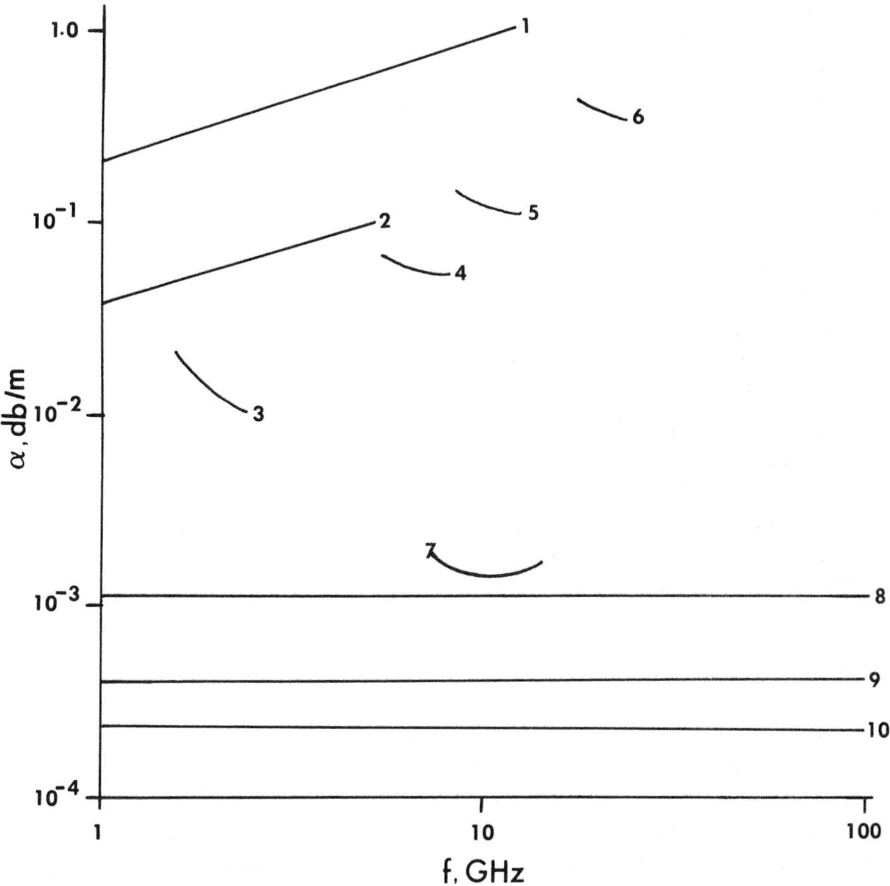

Figure 1.11 Attenuation (dB/m) versus frequency (GHz) for coaxial cables, waveguides (rectangular and circular), and optical fibers (multimode and single-mode): (1) 6.350-mm (1/4″) foam coaxial cable; (2) 22.225-mm (7/8″) air coaxial cable; (3) WR 340; (4) WR137; (5) WR90; (6) WR42; (7) 50.8-mm-diameter circular (copper) waveguide; (8) multimode-graded fiber (1300 nm); (9) single-mode fiber (1300 nm); (10) single-mode fiber (1500 nm).

1.4 APPLICATIONS AND REASONS

Fiber optic assemblies are finding their way into a wide range of applications, including:

- Phased array scanning;
- Locating signal processing centers remotely from antennas;
- Sequential scanning of multi-element antenna systems;

- Missile microwave guidance with antennas;
- Spectrum analyzers;
- Signal processing;
- Transmission systems for communication systems;
- Beam-forming networks for antennas used in radar, telecommunications, and satellites.

The main reasons for the above applications are:

- Low loss for all microwave frequencies;
- Small size;
- Light weight;
- Immunity to EMP, electrical noise, and RFI;
- Made of dielectric, therefore nonconductive;
- High flexibility and easy to arrange in a large network;
- No radiation loss;
- Large bandwidth compared to other transmission lines;
- Can operate in a higher temperature environment than waveguide, microstrip, stripline, coaxial cable, or TEM lines;
- Can be installed with a bend radius of only a few centimeters.

1.5 CONCLUSIONS

This chapter describes the background and history of optical fiber technology. A systematic table has been produced on the historical development of the optical fiber components and sources from 1850 to 1994 (see the Appendix).

The attenuation is an important parameter in deciding the type of transmission lines for the beamforming network in array antennas. Five transmission line technologies have been discussed and compared. It has been shown that the loss in the optical fiber transmission line is less than 0.0002 dB/m. At present, TEM line technology is used for the beamforming network in satellite array antennas, which is more lossy than optical fiber. Therefore, this technology opens a new door for the use of optical fiber in a beamforming network for a satellite antenna.

References

[1] Kompfer, R., "Optics at Bell Laboratories—Optical Communications," *Appl. Opt.*, Vol. 11, November 1972, pp. 2412–2425.

[2] Hondros, D., and P. Debye, "Elektromagnetische Wellen an Dielektrischen Drahten," *Ann. Phys.*, Vol. 32, 1910, pp. 465–476.

[3] Nathan, M. I., W. P. Dumke, G. Burns, F. H. Dill, Jr., and G. Lasher, "Stimulated Emission of Radiation from GaAs p-n Junctions," *Appl. Phys. Lett.*, Vol. 1, 1962, pp. 62–64.

[4] Quist, T. M., R. H. Radiker, R. J. Keys, W. E. Krag, B. Lax, A. L. McWhorter, and H. J. Zeigler, "Semiconductor Maser of GaAs," *Appl. Phys. Lett.,* Vol. 1, 1962, pp. 91–92.

[5] Hall, R. N., G. E. Fenner, J. D. Kinsley, T. J. Soltys, and R. O. Carlson, "Coherent Light Emission from GaAs Junctions," *Phys. Rev. Lett.,* Vol. 9, 1962, pp. 366–368.

[6] Holonyak, N, Jr., and S. F. Bevacqua, "Coherent (Visible) Light Emission from Ga(As1-xPx) Junction," *Appl. Phys. Lett.,* Vol. 1, 1962, pp. 82–83.

[7] Kroemer, H., "A Proposed Class of Hetrojunction Injection Laser," *Proc. IEEE,* Vol. 51, 1963, pp. 1782–1783.

[8] Kressel, H., and F. Z. Hawrylo, "Stimulated Emission at 300° K and Simultaneous Lasing at Two Wavelengths in Epitaxial Al_x Ga_{1-x} as Injection Length," *Proc. IEEE,* Vol. 56, 1968, pp. 1598–1599.

[9] Rupprecht, H., J. M. Woodall, G. D. Pettit, J. W. Crowe, and H. F. Quinn, "Simulated Emission from Ga_{1-x} Al_x As Diodes at 77° K," *J. Quant. Electron.,* Vol. QE4, 1968, pp. 35–42.

[10] Susaki, W., T. Sago, and T. J. Oku, "Lasing Action in (Ga_{1-x} Al_x) as Diodes," *J. Quant. Electron.,* Vol. QE4, 1968, pp. 422–427.

[11] Nelson, H., and H. Kressel, "Improved Red and Infrared Light Emitting Al_x G_{1-x} As Laser Diodes Using the Close-Confinement Structure," *Appl. Phys. Lett.,* Vol. 15, 1969, pp. 7–18.

[12] Kao, K. C., and G. A. Hockham, "Dielectric-Fiber Surface Waveguides for Optical Frequencies," *Proc. IEE,* Vol. 113, July 1966, pp. 1151–1158.

[13] Goodwin, A. R., W. O. Bourne, M. Pion, and P. R. Selway, "Fabrication and Performance Charactristics of C. W. Stripe Geometry GaAs/(GaA1) As Lasers for Optical Communications," *Proc. Electro-Optics/ Laser International Conference '76 UK,* Brighton, U.K., March 1976, IPC Science and Technology Press, pp. 4–7.

[14] Alferov, Z. I., "Possible Development of a Rectifier for Very High Current Densities on the Basis of a p-i-n (P-n-n^+, n-p-p^+) Structure with Hetrojunctions," *Soviet Physics-Semicon.,* Vol. 1, No. 3, September 1967, pp. 358–360.

[15] Ikegami, T., "Survey of Telecommunications Applications of Quantum Electronics—Progress with Optical Fiber Communications", *Proc. IEEE,* Vol. 80, No. 3, March 1992, pp. 411–419.

[16] Wedding, B., "Breakthrough in Multigigabit Systems Technology Due to Optical Amplifiers," *Tech. Dig. Optical Amplifiers and Their Applications,* Paper FC4, Santa Fe, NM, 1992, pp. 196–199.

[17] Kumar, A., "Future Communications Systems", *Microwaves & RF,* Vol. 26, No. 2, February 1987, p. 65.

[18] Thiennot, J., F. Pirio, and J. B. Thomine, "Optical Undersea Cable Systems Trends," *Proc. IEEE,* Vol. 81, No. 11, November 1993, pp. 1610–1623.

[19] Huxley, L. G. H., *A Survey of the Principles and Practice of Waveguides,* Cambridge: The University Press, and New York: The Macmillan Co., 1947.

[20] Ramo, S., J. R. Whinnery, and T. Van Duzer, *Fields and Waves in Communication Electronics,* New York: John Wiley and Sons, 1965.

[21] Jordan, E. C., and K. G. Balmain, *Electromagnetic Waves and Radiating Systems,* Englewood Cliffs, NJ: Prentice-Hall, 1968.

[22] Pozar, D. M., *Antenna Design Using Personal Computers,* Norwood, MA: Artech House.

[23] Gupta, K. C., R. Garg, and I. J. Bahl, *Microstrip Lines and Slotlines,* Dedham, MA: Artech House, 1979.

[24] Bahl, I. J., and R. Garg, "A Designer's Guide to Stripline and Circuits," *Microwaves,* Vol. 17, No. 1, January 1978, pp. 90–96.

[25] Kumar, A., "Effect of Nuclear Radiation on Microstrip Antennas," *Proc. JINA 1988,* Nice, France, November 1988.

[26] Kumar, A., *Effect of Nuclear Radiation on Materials in Space,* Research report 1989, AK Electromagnetique Inc., Quebec, Canada, June 1989.

[27] Moriyama, T., O. Fukuda, K. Sanada, K. Inada, T. Edahiro, and K. Chida, "Ultimately Low OH Content VAD Optical Fiber," *Electron. Lett.,* Vol. 16, August 1980, pp. 699–700.

[28] Bagley, B. C., C. R. Kurkjian, J. W. Mitchell, G. E. Peterson, and A. R. Tynes, "Materials, Properties, and

Choices,'' *Optical Fiber Telecommunications* (S. E. Miller and A. G. Chynoweth, eds.), New York: Academic Press, 1979.

[29] Kaiser, P., and D. B. Keck, ''Fiber Types and Their Status,'' *Optical Fiber Communications* II (S. E. Miller and I. P. Kaminow, eds.), New York: Academic Press, 1988.

[30] Miya, V., Y. Terunuma, T. Hosaka, and T. Miyashita, ''Ultra Low Loss Single-Mode Fibers at 1.55 μm,'' *Electron. Lett.,* Vol. 15, 1979, pp. 106–108.

[31] Nagel, S. R., J. B. MacChesney, and K. L. Walker, ''An Overview of the MCVD Process and Performance,'' *IEEE J. Quantum Electron.,* Vol. QE-18, April 1982, pp. 459–476.

[32] Osanai, H., T. Shioda, T. Mariyama, S. Araki, M. Horiguchi, T. Izawa, and H. Takata, ''Effects of Dopants on Transmission Loss of Low OH Content Optical Fibers,'' *Electron. Lett.,* Vol. 12, October 1976, pp. 549–550.

[33] Olshansky, R., ''Propagation in Glass Optical Waveguides,'' *Rev. Mod. Phys.,* Vol. 51, April 1979, pp. 341–367.

[34] Maurer, R., ''Glass Fibers for Optical Communications,'' *Proc. IEEE,* Vol. 61, April 1973, pp. 452–462.

[35] Keck, D. B., ''Fundamentals of Optical Waveguide Fibers,'' *IEEE Comm. Magazine,* Vol. 23, May 1985, pp. 17–22.

[36] Andreansen, S. B., ''New Bending Loss Formula Explaining Bends on Loss Curve,'' *Electron. Lett.,* Vol. 23, No. 21, October 1987, pp. 1138–1139.

[37] Glodis, P. F., C. H. Gartside III, and J. S. Nobles, ''Bending Resistance in Single mode fibers,'' *Proc. OFC/IOOC'87,* Reno, Nevada, 1987, Paper TUA3, p. 41.

[38] Marcuse, D., ''Curvature Loss Formula for Optical Fibers,'' *J. Opt. Soc. Amer.,* Vol. 66, 1976, pp. 216–220.

[39] Gloge, D., ''Propagation Effects in Optical Fibers,'' *IEEE Trans. Microwave Theory Tech.,* Vol. MTT-23, January 1975, pp. 106–120.

Chapter 2

Optical Fiber Types, Fabrication, Radiation, and Reliability

2.1 INTRODUCTION

The attenuation in an antenna feed network is an important parameter in deciding the type of transmission line. It is shown in Chapter 1 that the optical fiber has the lowest loss of all presently available transmission lines (waveguides, coaxial cables, microstrip lines, and striplines). This chapter provides descriptions of fiber types including advantages between single-mode and multimode fibers, fabrication techniques, effects of nuclear radiation, and life-time strength and fatigue for optical fibers.

2.2 FIBER TYPES

There are two types of fiber that are commercially available for signal transmission. They are called:

- Single-mode (SM);
- Multimode (MM).

Multimode fiber is further classified as:

- Step-index multimode (SIMM);
- Graded-index multimode (GIMM).

2.2.1 Single-Mode Fiber

A single-mode fiber transmits light in a single waveguide mode for all wavelengths longer than the cut-off wavelength (λ_c); however, it contains two or more lowest order modes with orthogonal linear polarizations at shorter wavelengths. The cut-off wavelength is calculated from [1–3]

$$\lambda_c = \frac{2\pi a n\sqrt{2\Delta}}{2.405} \tag{2.1}$$

where a is the core radius of the fiber, n is a homogeneous core refractive index, and Δ is a fractional index difference between core and cladding $\Delta = \Delta n/n$.

Figure 2.1(a) shows the structural parameters in terms of outside diameter and core diameter. It contains a core of high-index glass surrounded by a much thicker cladding of lower index glass, separated by an abrupt boundary. Radial refractive index profiles for several special single-mode fibers are shown in Figure 2.1(b). The peak core index is shown by $n(1 + \Delta)$ in the figure. A single-mode fiber provides very broad bandwidth, and the aperture is smaller than other types of fiber. The core diameter of a typical single-mode fiber is approximately 8 μm for operation at 1.3 μm or 1.5 μm. The thin core design leads to tight mechanical tolerances in coupling with other fibers.

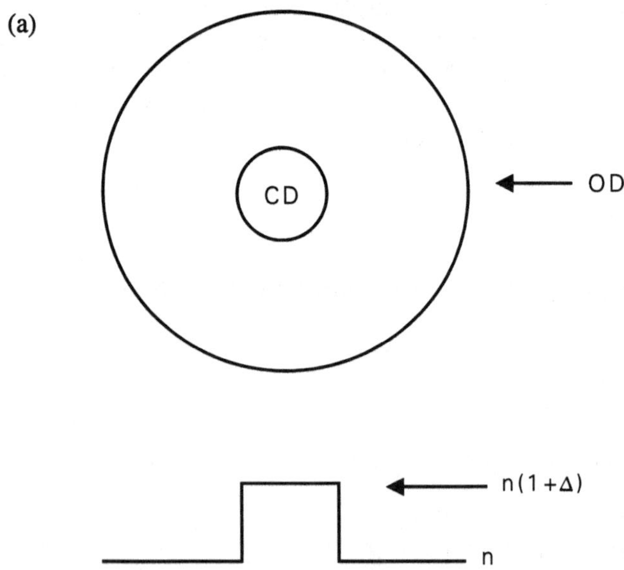

Figure 2.1 Single-mode (a) fiber parameters and (b) fiber index profile (OD = outside diameter; CD = core diameter).

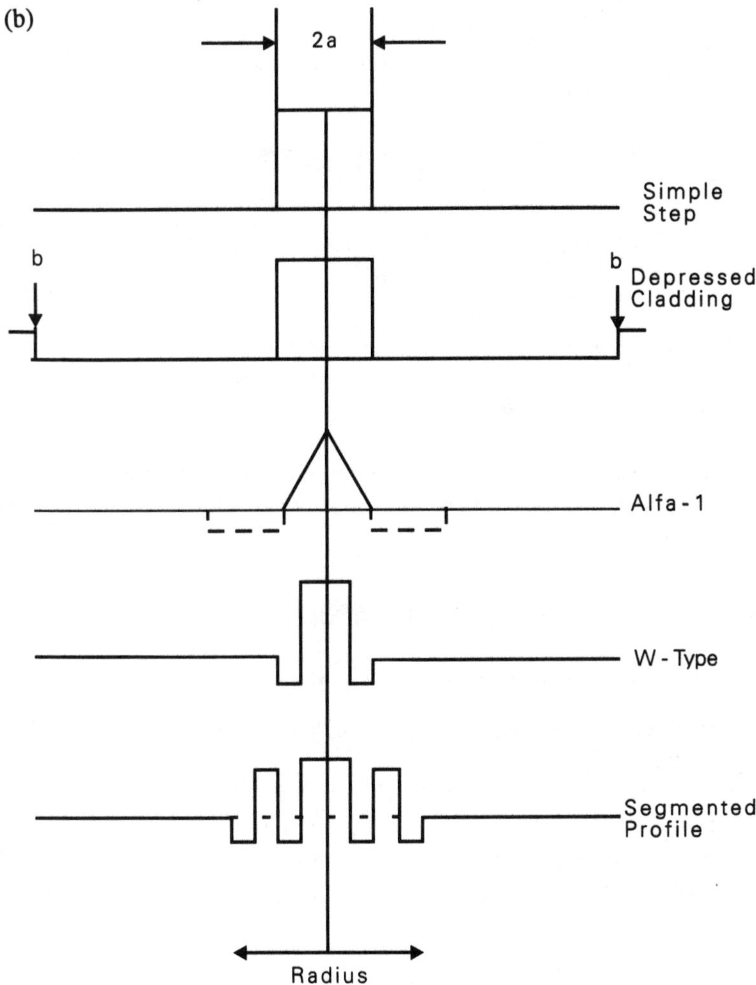

Figure 2.1 (continued)

2.2.2 Step-Index Multimode Fibers

The propagation of light rays in the step-index multimode fiber follows a zig-zag path (Figure 2.2) and contains many modes. The core diameter of fiber is in the range of 50 μm to 100 μm, and the outer clad diameter varies in the range of 125 μm to 140 μm.

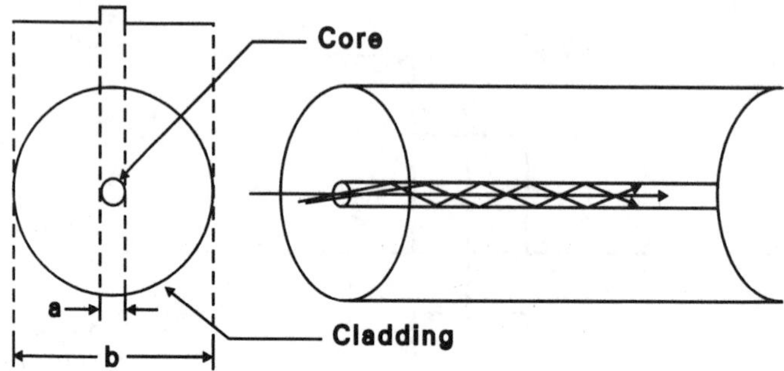

Figure 2.2 Refractive index profile of the step index fiber (a = 50 μm to 100 μm and b = 125 μm to 140 μm).

2.2.3 Graded-Index Multimode Fibers

Figure 2.3 shows a sketch of a graded-index multimode fiber. The path of propagation of light rays is sinusoidal, and the core and the outside diameter of clad are the same as for the step-index fiber.

2.3 ADVANTAGES OF SM FIBER OVER MM FIBER

SM fiber has the following advantages over MM fiber.

- *Attenuation loss:* It is shown in Figure 1.11 (Chapter 1) that SM fiber has lower fiber attanuation than MM fiber.
- *Rayleigh scattering coefficient:* Rayleigh scattering is caused by the interaction of light and the granular appearance of atoms and molecules on a microscopic scale. There is a lower Rayleigh scattering coefficient in SM fiber. A typical comparison is shown in Table 2.1.

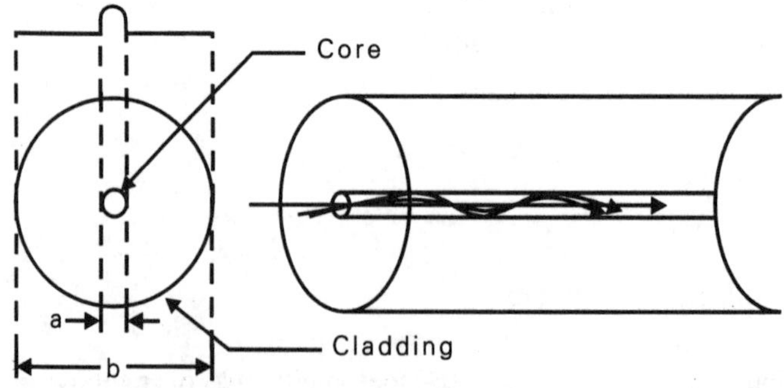

Figure 2.3 Refractive index profile of the graded index fiber (a = 50 μm to 100 μm and b = 125 μm to 140 μm).

Table 2.1
Scattering Coefficients

Fiber type	Wavelength (μm)		
	0.85	1.3	1.55
MM fiber	2.2 to 3	0.44 to 1	0.25 to 0.8
SM fiber	1.82 to 2.5 (double-mode)	0.27 to 0.6	0.15 to 0.4

- *Nuclear radiation for application in antennas:* SM fiber has better resistance to nuclear radiation than MM fiber.
- SM fiber has lower joint and coupler losses.
- SM fiber has lower dispersion/higher bandwidth.
- Preservation of polarization and coherence cannot be accomplished with MM fiber for distances exceeding a few or a few tens of meters.
- SM fiber has lower production cost than MM fiber.

Despite the enormous advantages of SM fiber over MM fiber, there are a few points that should be kept in mind with SM fiber systems. They have small mode diameter (9 μm to 10 μm at 1300 nm), which makes fiber-to-fiber joining more difficult, although lower loss values are attainable if due care is taken.

2.4 MATERIALS IN THE FABRICATION

Various materials are suitable for fabricating infrared fibers. The most promising fiber material is glass; and it can be manufactured from various metal oxides, metal halides, and chalcogenides. The transmission loss in silica fiber was more than 0.3 dB/m in 1969, and today the loss is less than 0.0002 dB/m. The transmission loss has been reduced on other fibers, which are manufactured using heavy-metal oxides, fluoride glasses, chalcogenide glasses, and halide crystals since 1980. Silica (SiO_2) has superior transmission and mechanical properties compared with other materials. Germanium oxide (GeO_2) is good for ultra low loss transmission at longer wavelengths in the infrared compared with silica. This is because the infrared absorption due to germanium and oxygen lattice vibrations is shifted toward the red owing to the heavier weight of germanium. The effect of other metal oxide glass dopants on the refractive index has been described by Yoshida [4] as shown in Figure 2.4. In the figure, the refractive index (n) is plotted on the vertical axis and the horizontal axis represents the concentration F, GeO_2, P_2O_5, B_2O_3, F (Mole %) in the fiber.

Baldwin et al. [5] published a review paper on metal halides in 1981. These materials incorporate an ion of the halogens (Cl_2, F_2) in place of oxygen. Poulain et al. [6]

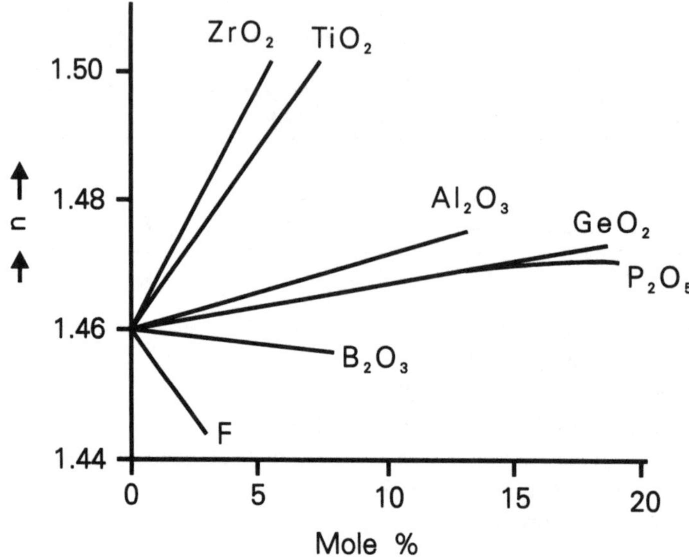

Figure 2.4 Index modifying effect of GeO_2, P_2O_5, B_2O_3, *F*, and other common dopant for high-silica fibers (n = refractive index).

reported fluorozirconate glasses in 1975. The application of multicomponent fluoroberlatte glasses to manufacture low-loss infrared fibers was developed at Corning. Ohsawa et al. [7] reported ZrF_4-BaF_2-LaF_3-AlF_3-NaF (ZBLAN) material for fiber glass in 1981.

Chalcogenide glasses are manufactured from metals such as arsenic, germanium, and antimony combined with the heavier elements in the oxygen family like S, Se, and Te. The electronic absorption edges of these glasses are in the middle infrared, visible, or near-infrared region. These materials provide high reflective index and have low glass transition temperature in the range of about 200° to 300° C. Chalcogenide glasses are manufactured by mixing and melting purified elemental materials in vacuum. One demerit of these materials is that they are damaged easily by moisture. Kanamori et al. [8] showed that these materials provide a weak intrinsic absorption tail at around 5 μm that limits the maximum attainable attenuation to 0.001 dB/m to 0.01 dB/m. However, they are very attractive for the application of short fiber applications in the infrared region.

Shimizu et al. [9] described the application of fiber cores doped with rare-earth dopants like erbium. It is shown in their paper that fiber cores doped with erbium can be used to amplify optical signals at 1.55-μm wavelength. Therefore, a keen interest has been shown in doping silicates or fluoride glasses with rare-earth ions like erbium, neodymium, and praesodymium.

2.5 FABRICATION TECHNIQUES

Several fabrication techniques have demonstrated excellent manufacturability. These are summarized as:

- Vapor-phase axial deposition (VAD);
- Outside vapor deposition (OVD);
- Modified chemical vapor deposition (MCVD);
- Plasma chemical vapor deposition (PCVD).

2.5.1 Vapor-Phase Axial Deposition (VAD)

This technique was reported by Izawa et al. [10] in 1977. Fabrication techniques for the various kinds of VAD fibers such as graded-index fibers, single-mode fibers, and single polarization fibers have been developed and improved. A schematic diagram of the VAD technique for the fabrication of fibers is shown in Figure 2.5. A typical VAD process is summarized in Table 2.2 [11], which shows that soot preform is the first process of VAD. In this process, fine glass particles synthesized in the oxy-hydrogen flame are deposited onto the end surface of the seed rod, which is made of silica. A porous preform is then grown along the axial direction.

The end position of the porous preform is constantly regulated by monitoring it with

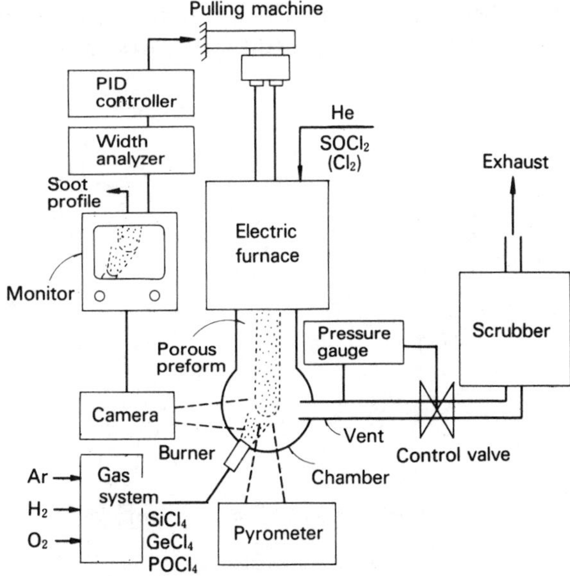

Figure 2.5 Schematic diagram of VAD apparatus. (© 1982 IEEE.)

Table 2.2
Typical VAD Process

Soot Preform	Deposition of fine glass particles by flame hydrolysis reaction.
Dehydration treatment and sintering	Dehydration by Cl_2 or $SOCl_2$ at about 1200° C then sintering at about 1500° C. The rod diameter after the sintering is about 25 nm.
Preform elongation	In an oxy-hydrogen flame, elongated rod diameter is about 10 nm.
Jacketing	Thick wall silica tube with rod-in-tube method.

Source: [11].

an internal television camera for a uniform growth of the preform. The second process is called dehydration treatment and sintering. In this process the dehydration of the VAD fiber is achieved by increasing the $SOCl_2$ vapor in oxygen gas [12–16] or chlorine gas treatment in the furnace. Active Cl ions produced by the thermal decomposition of $SOCl_2$ react with OH ions and H_2O molecules as described by Sudo et al. [12]:

$$H_2O + SOCl_2 \rightarrow SO_2 + 2HCl$$
$$\text{Si-OH} + SOCl_2 \rightarrow \text{Si-Cl} + SO_2 + HCl$$

Chida et al. [14] showed in Figure 2.6 that there are two critical temperatures of 700° and 1200°C for the above chemical reactions. The chemical reaction between OH ions and $SOCl_2$ happens on the particle surfaces rather than OH diffusion. The main reason for this

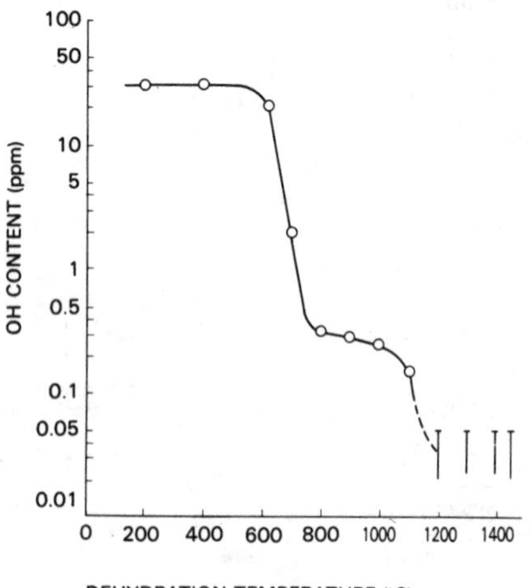

Figure 2.6 OH-content versus the dehydration temperature. (© 1982 IEEE.)

is that the OH diffusion length in silica glass during the dehydration time is approximately 5.5 μm at 600°C, which is larger than the particle size. The residual OH concentration is also a function of the $SOCl_2$ flow rate. Edahiro et al. [15] showed that the larger the flow rate, the smaller the residual OH content. It is also found out that a dehydration temperature above 1200°C is required to reduce the OH content below 0.1 ppm.

The third process is called consolidation. In the continuous consolidation process, the pulling speed of the consolidated preform must be matched with the relation among the soot preform growing speed, shrinkage rate of the soot preform in the axial direction, and the elongation rate of transparent preform at the necking down position, which is the function of weight of soot preform, bulk density of soot preform, dopant content, and sintering temperature.

The fourth process is called preform elongation, which is performed in an oxy-hydrogen flame. The final process is the jacketing silica tube.

For the mass production of optical fiber, it is desirable to dehydrate the soot preforms simultaneously with the consolidation process. Chida et al. [14] developed the technique using a specially designed muffle-type resistance furnace as shown in Figure 2.7.

Since 1985, the VAD fiber is under mass production for the medium/small-capacity optical fiber transmission system. The VAD method is suitable for the single-mode fiber because of low OH content and large preform size. In Japan, the VAD single-mode fiber is used for high-capacity optical fiber transmission systems in land and under-sea areas.

Tamura et al. [17] described the fluorine-doped dispersion-shifted fiber. The fluorine doping has been useful in the reduction of transmission loss in VAD single-mode fibers. A high yield of low-loss fiber under 0.0002 dB/m at 1.55 μm was achieved in fully fluorine-doped single-mode fibers with nonshifted design in which fluorine is doped in cladding and is slightly added to the pure silica core as described by Ogai et al. [18]. Fluorine doping is achieved in the VAD process by heating porous soot preform in a fluorine gas atmosphere. The reduction of the refractive index is adjusted by the gas pressure. Ogai et al. [18–20] described the characteristics of fluorine-doped silica fibers. The strength of the fluorine-doped fiber is reduced by 10 percent. From the tensile test, almost the same level of reduction of Young's modulus was confirmed.

Sakaguchi et al. [21], Katuyama et al. [22], and Miyamoto et al. [23] developed VAD single-mode fibers with ultimate low-core eccentricity. Using the VAD process a core and a whole cladding were synthesized with a high-deposition rate.

Wada et al. [24, 25] and Sudo [26] reported the high-speed and large-scale preform fabrication by the two-step VAD process. The two-step consolidation processes are as follows: (1) glass particles of a soot preform are stuck to one another, but the interstices between the glass particles are linked together, and gas in the preform is diffused through the passage; (2) as sintering goes on, the passage is cut to pieces, and a number of isolated pores are formed in the preform. Residual gas in the bubbles can no longer exit through the passage. The final consolidation is achieved by gas diffusion in the bubbles into the glass body.

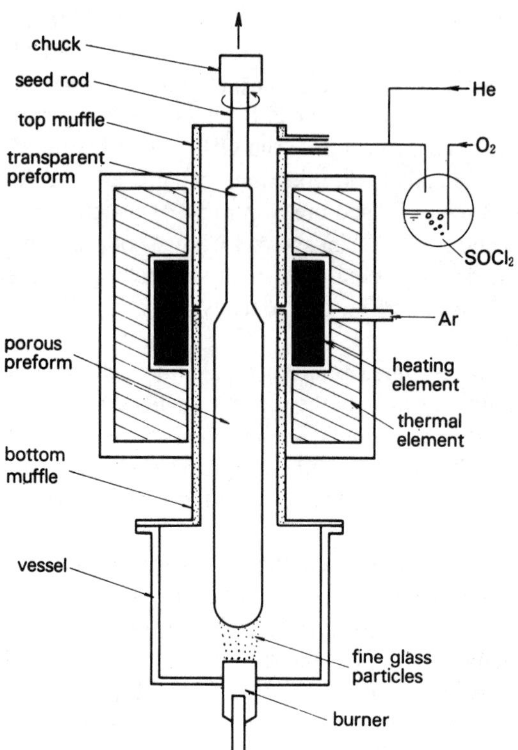

Figure 2.7 Muffle-type electric furnace for consolidation simultaneously with dehydration in $SOCl_2$ atmosphere. (© 1982 IEEE.)

In 1993, the Japanese delegation [27, 28] reported that they were developing antenna beamforming networks using VAD single-mode optical fiber for land and space applications.

2.5.2 Outside Vapor Deposition (OVD)

The outside vapor-phase oxidation and deposition process, commonly called the OVD process, has been shown to be a very flexible and economic process for fabricating optical fibers. Raw material for the fiber used in this method is a high silicon-containing liquid material with a high vapor pressure at a temperature above ambient. Silicon tetrachloride ($SiCl_4$) is a low-cost material that meets the high vapor pressure and purity characteristics. The index of refraction of silica glass can be increased by adding metal oxide dopants such as Germanium oxide (GeO_2), Germanium tetrachloride ($GeCl_4$), P_2O_5, B_2O_3, and TiO_2.

The OVD method uses the above chemicals which are metered into the deposition burner through a delivery system that provides precise control of flows, mixing of components, and vaporization. There are various types of vapor generation systems such as bubblers, direct chemical vaporization apparatus with a gas flow controller, and a metering pump with a subsequent vaporization chamber. Blankenship et al. [29], Schultz [30], Blankenship [31], Bailey et al. [32], and Blankenship et al. [33] described a standard bubbler vapor generation system that employs a mass flow controller to provide precise control of the carrier gas (oxygen or nitrogen). In Figure 2.8, the carrier gas bubbles through $SiCl_4$, $GeCl_4$, and $PoCl_4$ in the container, thereby vaporizing some of the chemical, and then passes to the deposition burner. Figure 2.9 shows a typical deposition burner including a soot preform chamber. The deposition burner contains a series of concentric orifices. The center orifice is for metal halide vapors; an adjacent orifice is used for a shielding gas that prevents premature reaction of the chlorides; and an outer orifice is used

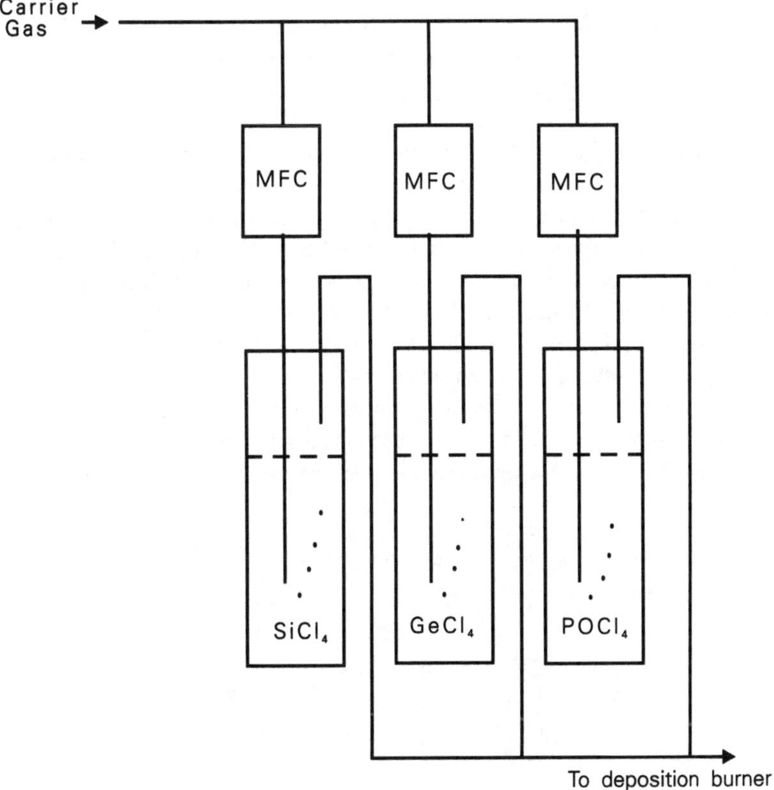

Figure 2.8 Bubbler vapor generation system. (© 1982 IEEE.)

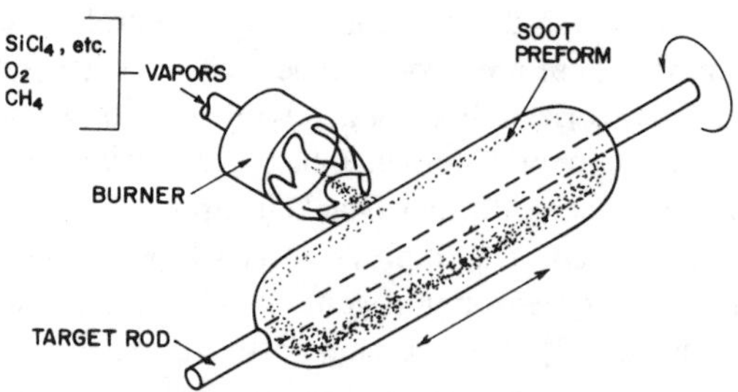

Figure 2.9 Soot deposition process. (© 1982 IEEE.)

for the fuel gas and oxygen mixture. The heat of the flame, along with the oxygen, causes the metal chloride vapors to react, forming tiny spheres of metal-oxide material. These spheres are directed toward a target mandrel or soot preform where a fraction of the glass is collected. Figure 2.9 shows the soot depostion process. Figure 2.10(a) shows sintering of the porous preform to a dense glass preform. It is done by supporting the soot preform from one end and passing it vertically through a hot zone of about 1500°C inside a refractory muffle furnace. By using helium with a few percent chlorine as the atmosphere within the muffle, the glass preform is effectively purged of hydroxyl ions.

Figure 2.10(b) shows the method of drawing glass preforms into fiber. The process uses induction of a resistance furnace at about 2000°C. The glass preform is fed into the tube that is inside the furnace, and the free-formed glass fiber is pulled out of the bottom. The glass fiber is coated with protective polymers to retain its intrinsically high strength.

The key areas of advancement during the past ten years are in fiber design, dopants, fiber quality, and rate. Progress has also been made in polarization-retaining single-mode fibers that have low attenuation even when coiled to diameters less than 25 mm. In the area of new dopants, tests have shown cerium-doped OVD fiber to have significantly improved radiation resistance. The dopant used is an organometallic designed specifically for the process. The use of boron in graded-index multicore fiber has made it possible to

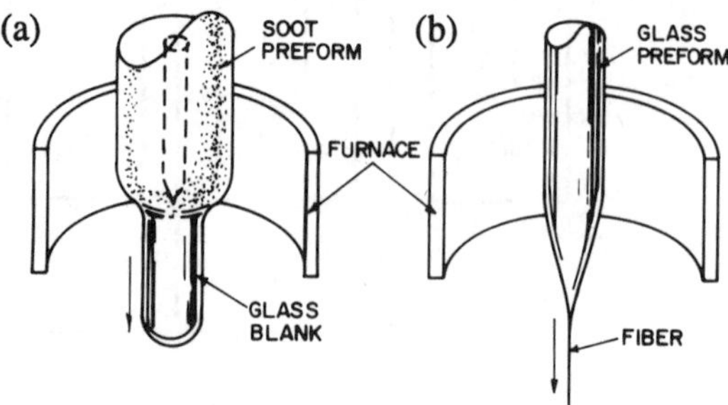

Figure 2.10 (a) Sintering of the porous preform; (b) method of drawing of glass preform into fiber. (© 1982 IEEE.)

produce fiber that has an attenuation at 1300 nm of 0.0009 dB/m. In the area of the fiber quality, advances have been made in multimode and low-loss fibers using fluorine [34]. Fiber strength has improved to the point where 150-kpsi to 200-kpsi fibers are available for highly stressed systems such as feed networks in large ground and satellite antennas.

2.5.3 Modified Chemical Vapor Deposition (MCVD)

Chemical vapor deposition of glasses has been used in the chemical industry since the late 1950s. To increase the deposition rate, scientists [35–50] have used a homogeneous reaction at high temperatures with high-reactant concentrations, and the process is called the modified chemical vapor deposition (MCVD). The process injects vapors of materials like $SiCl_4$, $POCl_3$, $GeCl_4$, BCl_3, SiF_4, SF_6, Cl_2, and freon inside a silica tube that is heated externally by an oxyhydrogen burner to produce glass. This process is also known as the inside vapor deposition process (IVD). Figures 2.11 and 2.12 show schematics of the process to collect vapor. In the figure, silica and P_2O_5 are derived by direct oxidation of $SiCl_4$ and $POCl_3$ vapors. However, we can add, for example, $GeCl_4$, BCl_3, SiF_4, SF_4, and we can collect corresponding vapors by bubbling oxygen through temperature-controlled Dreschel bottles containg the liquid materials. Within the silica-supporting tube, a dense dispersion of small glass particles is formed and is fused onto the walls of the tube at a temperature that is dependent on composition. In Figure 2.13, the P_2O_5 concentration of the glass is most easily controlled by holding the oxygen flow through the $SiCl_4$ constant and varying the oxygen flow through the $POCl_3$ by means of a mass-flow controller. The fused silica tube is heated to about 1400°C by the burner that travels the length of the tube. The fused silica tube is rotated with a constant speed. A new layer of glass is deposited at each pass of the burner. After certain layers (typically 50) of glass are formed on the silica tube, the temperature of the burner is raised so as to collapse the tube into a solid preform. Fiber is drawn from the preform as shown in Figure 2.14.

Since the invention of the MCVD technique in 1974, steady progress has been made in fiber performance, process improvements, and the mass fabrication of fiber by this method. In the mid-1980s Western Electric in the United States and many other companies throughout the world manufacture fibers using this method, and their fibers are used by various communication industries.

The sol-gel [47–49] method is used to apply thin doped silica layers inside MCVD preforms for the fabrication of rare-earth-doped fiber amplifiers. In this method, an MCVD substrate tube is filled with a doped alkoxide sol and drained at a constant rate. Upon collapse, the thin glass layer derived from the gel acts as a negligibly thin diffusion source of dopants. A typical MCVD cladding is dipcoated with a sol containing both rare-earth ions and index alternating dopants, such as germanium or aluminum. This process offers good reproducibility, allows localization of rare-earth ions within the MCVD core, and minimizes dopant clustering by delivering dopant ions already incorporated in a silica network.

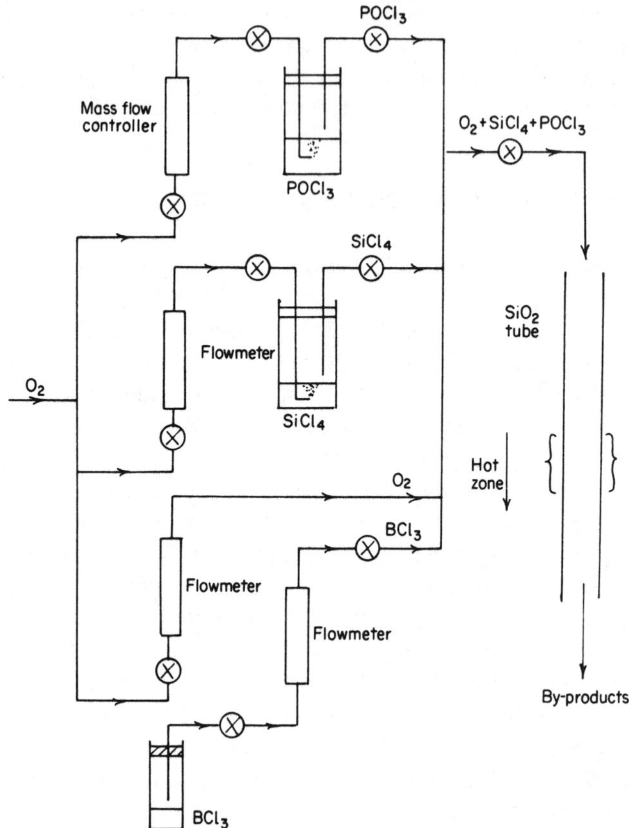

Figure 2.11 Vapor-collecting system.

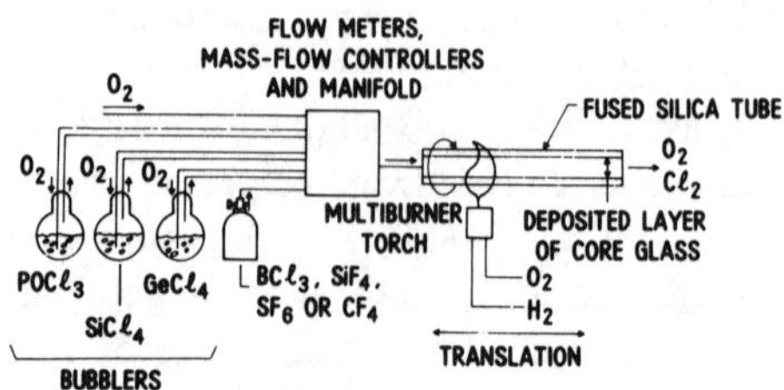

Figure 2.12 Schematic diagram of MCVD apparatus. (© 1982 IEEE.)

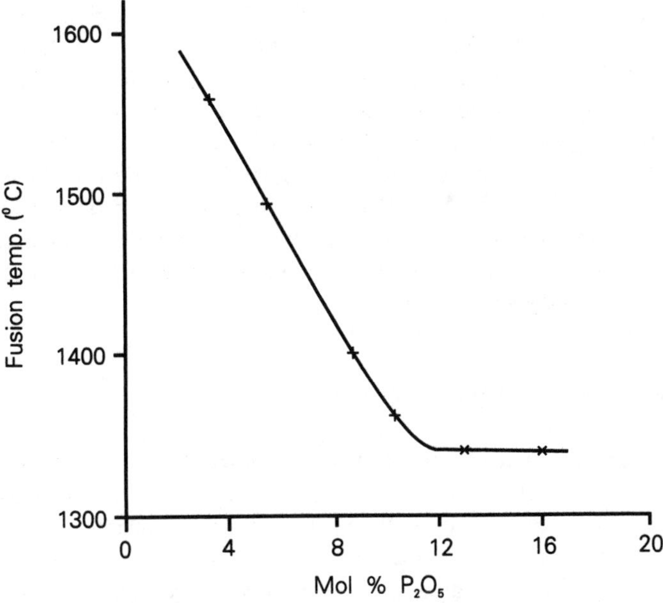

Figure 2.13 Glass fusion temperature as a function of P_2O_5 concentration.

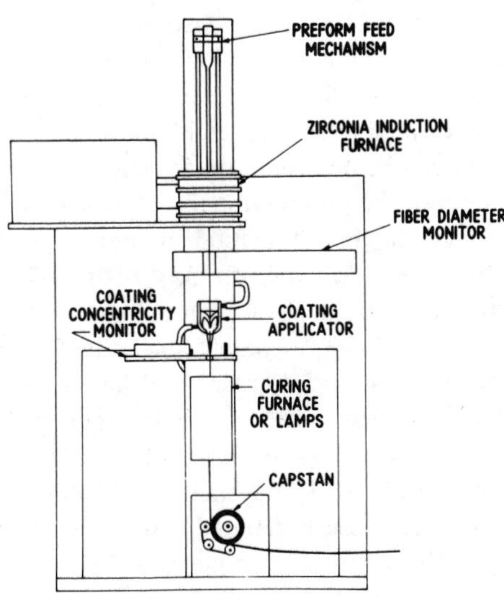

Figure 2.14 Schematic of state-of-the-art fiber-drawing apparatus. (© 1982 IEEE.)

A new technique has been reported [50] for the incorporation in MCVD of low-vapor-pressure compounds into the core of optical fibers. In this process an organic solvent and TEOS are used to dissolve organometallics as rare-earth acetylacetonates and aluminum or gallium butoxide. The liquid is then nebulized into a fine mist with a 1.5-MHz transducer, which produces aerosol particles with a size distribution peaked near 5 μm. This permits the convenient transport of the aerosol from the aerosol generator through a rotating seal into the MCVD substrate tube. The aerosol deposition technique permits a more convenient method for controlling the radial distribution of dopant. The aerosol combustion of organometallics is a new way of making glass with advantages over solution-doping techniques for fiber lasers.

2.5.4 Plasma Chemical Vapor Deposition (PCVD)

The process uses a plasma instead of flame to initiate chemical reaction in a tube to prepare fibers. Both microwave plasmas and RF plasmas have been studied by many researchers [51–58]. In this, it is difficult to achieve a deposition rate over 1 g/min because of low pressure and the nature of the process. The RF plasma process that operates at atmospheric pressure achieves high-reaction rates and efficiencies. Figure 2.15 shows a schematic of the plasma-enhanced modified chemical vapor deposition (PMCVD) process. This process uses an oxygen RF plasma that is operated at atmospheric pressure and is centered in the substrate tube with an annulus that is greater than 10 mm between the visible edge of the fireball and the tube. Fleming et al. [51] reported the deposition rate of silica up to 5 g/min. The fused layers deposited are typically 45 μm to 75 μm thick and have initial densities that are about 65 percent theoretical.

Kats [54] described the PCVD process using microwave energy. Figure 2.16 shows a schematic of the PCVD process. In this case a very high temperature is generated in the silica tube and a complete conversion of the metal halides into soot occurs. Typically, up to 2000 thin layers of glass can be deposited on the silica tube.

Bauch [55, 56] developed the plasma impulse chemical vapor deposition (PICVD) method, which is well suited for the fabrication of preform by inside deposition. The experimental set-up has been given in reference [55]. A first film of fluorine-doped SiO_2 and a second film of pure SiO_2 were deposited onto the inner side of the tube. After the deposition of pure SiO films with low OH content, the collapse of the tube was performed by several runs with an oxygen-hydrogen burner. If during the collapse process oxygen is passed through the tube, PICVD fibers with a pure SiO_2 core show a high attenuation in contrast to fibers with a doped SiO_2 core. Reducing the partial pressure of oxygen in the tube during the collapse process by passing a mixture of oxygen and helium through the tube causes the fiber loss to decrease remarkably. It is also found that if the collapse process is accomplished in an oxygenfree inert gas atmosphere, a very low loss (less than 0.0001 dB/m) fibers can be produced.

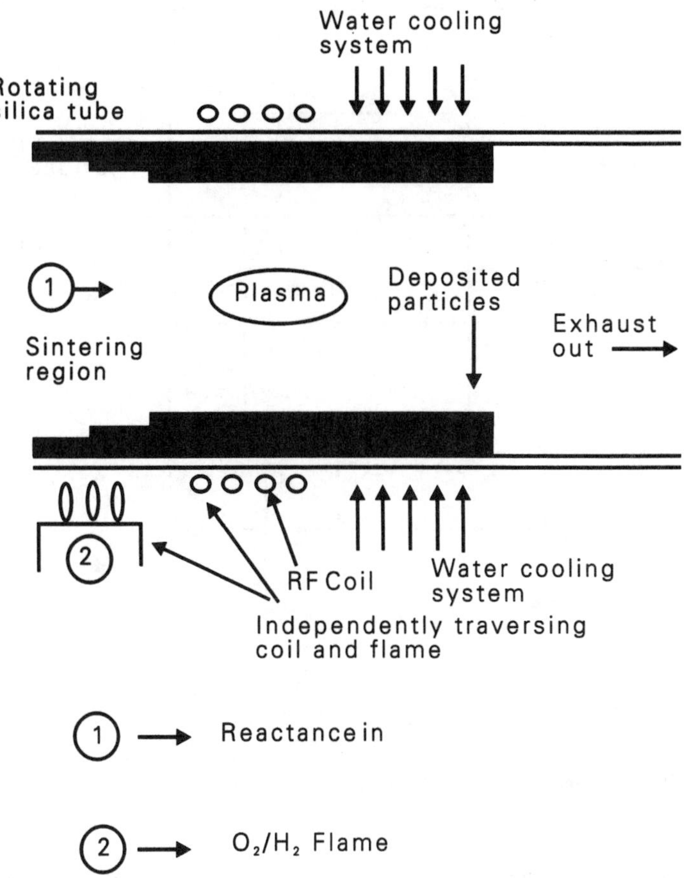

Figure 2.15 Schematic of the plasma-enhanced MCVD process.

2.6 NUCLEAR RADIATION-HARDENED FIBERS

Radiation-hardened optical fibers provide long life in the nuclear radiation environment (Earth and space applications) where a dose of radiation due to nuclear radiation is measurable. Nagasawa et al. [59], Wei et al. [60], Evans et al. [61], Oe et al. [62], and Huff et al. [63] examined hydrogen-doping of optical fibers to increase the nuclear radiation hardness. Miller et al. [64] reported a systematic study of the effects of hydrogen-doping on four types of single-mode optical fibers to improve their radiation hardness for a low-temperature (150°C) and 10^7 rad (high-dose) radiation dose in space environments. Table 2.3 provides the core size and operating wavelength values for fibers designated A to D. Fiber A has a silica core and fibers B to D have germania-doped cores and

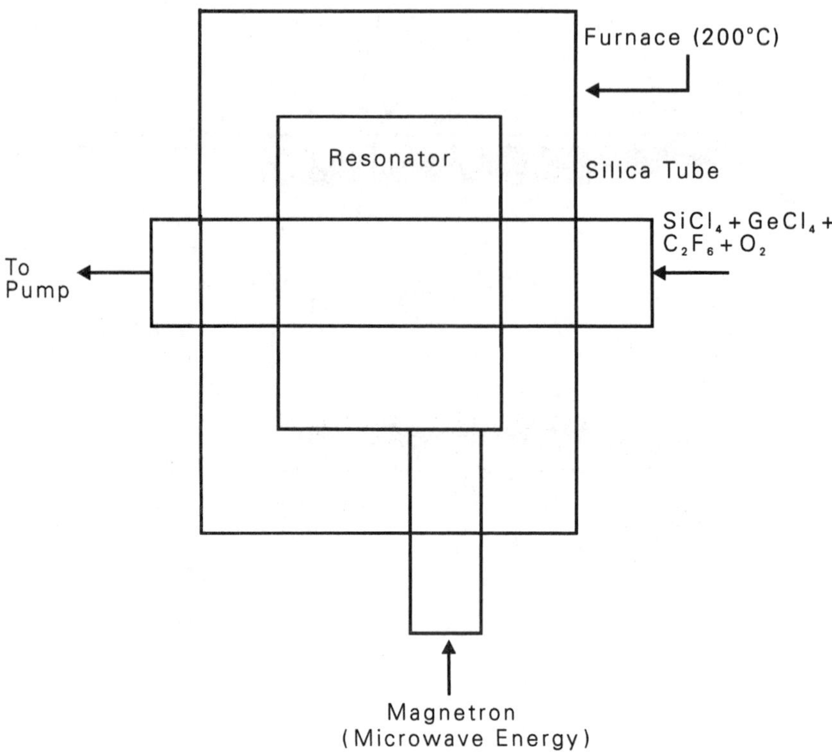

Figure 2.16 Schematic of PCVD process.

phosphorusfree inner claddings. Miller et al. [64] treated these fibers for three days at 50°C in ambients containing 0.1, 1.0, 10.0, and 100.0 atmospheres of hydrogen, resulting in 8.2, 82.0, 920.0, and 8200.0 ppm hydrogen in SiO_2, respectively. For a 1.7×10^5 rad dosage, reductions in radiation-induced losses of about one-half are obtained for fibers with a hydrogen concentration lower than 10 ppm. The added loss due to 10-ppm hydrogen is only 0.000018 dB/m at 1300 nm and 0.00005 dB/m at 1500 nm. Figure 2.17 plots radiation dose versus temperature of radiation-induced loss for hydrogen-doped hermetic fibers A, B, and C. Ref. [64] used fibers that were manufactured to permanently trap 2.7-ppm hydrogen using a hermetic carbon coating [63]. They reported that after applying 10^7 rad of radiation dose, the radiation-induced loss for fiber B decreases by 75 percent at −150°C and by 50 percent from −100°C to 20°C compared to untreated fibers. It is also reported that fibers A and C have reductions in loss by 50 percent and 15 percent, respectively. However, there is an increased loss of 6 percent in the case of fiber D. Among the three radiation-hardened fiber types, hydrogen-doped silica core fibers provide the widest operating range and smallest radiation-induced loss for high dose space appli-

Table 2.3
Core Size and Wavelength of Operation for Fibers A to D

Fiber Desination (%)	Core Size (μm)	Wavelength of Operation (nm)
A (0.35)	9.4	1.31
B (0.46)	13.2	1.31
C (0.86)	4.0	1.55
D (1.30)	2.9	1.31

cations. For the best trade-off between bending and radiation-induced losses, the fiber C (hydrogen-doped) is the best choice.

2.7 MECHANICAL RELIABILITY AND FATIGUE FOR SILICA GLASS FIBER

Mechanical life-time strength and fatigue are very important criteria in selecting antenna components. The criteria are at the top of the list if antennas are used in a satellite. In this section a short description is provided of the crack growth rate on the silica optical fibers.

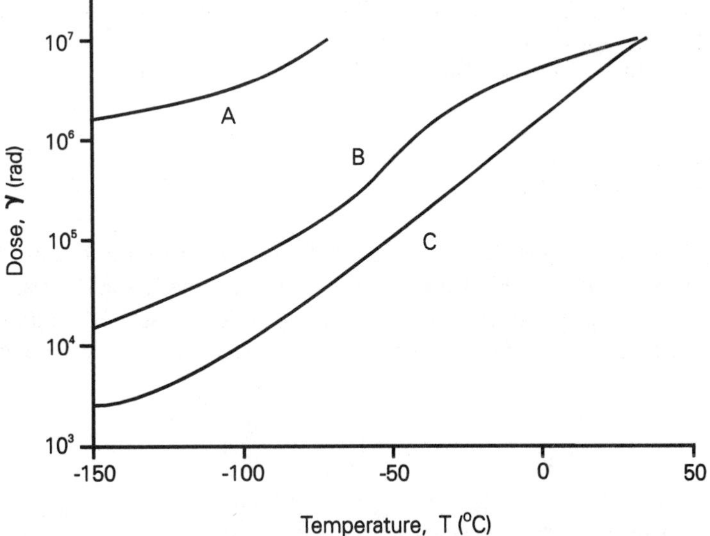

Figure 2.17 Plot of the radiation dose versus temperature of radiation-induced loss for hydrogen-doped hermetic fibers A, B, and C (radiation-induced loss <150 dB/km). (© 1992 OSA.)

The crack growth rate with time in the antenna beamforming network provides the actual figure of failure. In the case of satellite antennas, the crack growth rate limits the life of the satellite. During the launch, the satellite antennas pass through very high levels of vibration. It has been proven in laboratory experiments that the optical fiber can take more stress due to launching vibration than other beamforming networks such as waveguides, coaxial cables, microstrip lines, striplines, and other networks in antennas.

The life-time strength of silica optical fibers is limited by the moisture-enhanced stress corrosion cracking known as static fatigue. Nagel [65] reported the time to failure (T_s) for a fiber under static stress (σ_s) from the equation

$$T_s = B\, S_i^{n-2}\, \sigma_s^{-n} \tag{2.2}$$

where n and B are parameters characterizing static fatigue and S_i is the initial insert strength of fiber. The microcrack growth increases quickly under the presence of moisture. Kurkjian et al. [66] developed fiber sealing or ceramic jacketing to protect the fiber from moisture.

Krause et al. [67] and Zhurkov [68] reported that the time to failure for a fiber on static load is based on a thermofluctuation model. They used the fiber surface under the hermetic coating, and they assumed that there is no moisture on the fiber surface. In hermetically coated glass fibers, the growth of surface and internal defects is associated with the thermofluctuation mechanism. Therefore, the defect growth rate will be greater in those parts of the fiber that are affected by thermal stretching stresses due to the difference in the linear expansion. According to the thermofluctuation theory, the crack develops due to thermal fluctuations.

The equation used to calculate the time to failure is given by [67]

$$T_s = t_0 \exp\left[\frac{U_0 - b\sigma_s}{RT}\right] \tag{2.3}$$

where t_0 is the value approaching the period of the atom thermal oscillations (10^{-13}/s), U_0 is the stress-free activation Si-O bond energy (110 kcal/mol), static stress σ_s, b is the parameter determined by the value of activation volume and size of the initial defect, R is the gas constant, and T is the absolute temperature. Equation (2.3) can be rearranged as

$$T_s = t_0 \exp\left[\frac{U_0}{RT}\left(1 - \frac{\sigma_s}{S_{i0}}\right)\right] \tag{2.4}$$

where $S_{i0} = U_0/b$ is the fiber strength in the absence of thermofluctuations. Figure 2.18 shows the dependence of the time to failure (t_s) on the static stress at the temperature of liquid nitrogen (T = 77K) and room temperature (T = 300K).

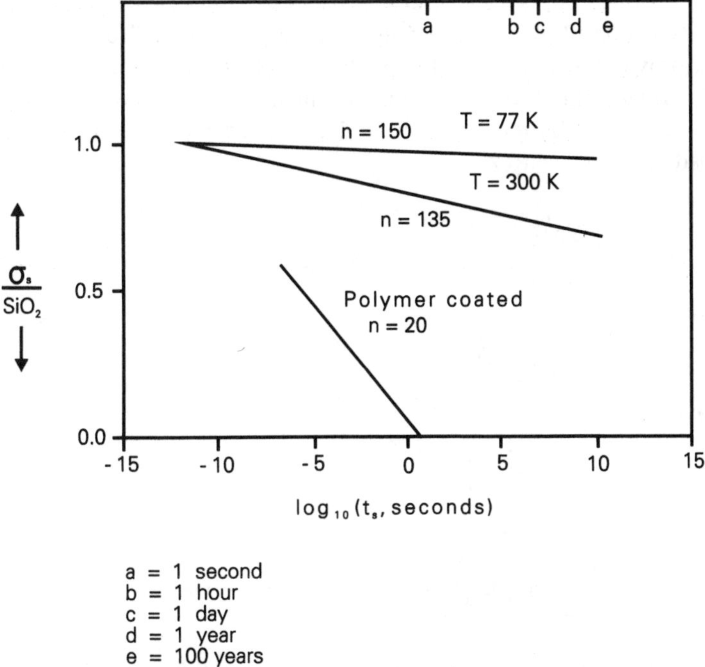

Figure 2.18 Static fatigue of hermetically coated silica fibers according to thermofluctuation theory. (© 1992 OSA.)

The slope is not constant when (2.4) is used, and the value of the parameter n in this case can be calculated from the expression [69]

$$n = -\frac{\partial \ln (t_s)}{\partial \ln (\sigma_s)} = \frac{U_0}{RT}\frac{\sigma_s}{S_{i0}}. \tag{2.5}$$

Bubel et al. [70] described the effect of constant stress on optical fibers. In 1992, Matthewson [71] presented models for the crack growth rate to the applied stress intensity. Usually, the mechanical reliability is determined by extrapolating from the results of accelerated static fatigue (crack-applied stress) tests using the subcritical crack growth model for fatigue which relates the time of failure (T_f) to the applied stress (σ_a)

$$T_f = 2\sigma_a^{-n}\frac{1}{AY^2(n-2)}\left(\frac{\sigma_i}{K_{Ic}}\right)^{n-2} \tag{2.6}$$

where A and n denote fatique constants determined by fitting to fatigue data, Y is a constant of order unity, K_{1c} is the critical stress intensity factor, and σ_i is the initial strength of material. Equation (2.6) is involved in life-time predictions. However, there are two sources that provide uncertainty in the prediction: (1) scatter in the accelerated data to which (2.6) is fitted, and (2) uncertainty in the kinetics of crack growth. Matthewson [71] considered four models for the crack growth kinetics, which relate to crack growth rate $\dot{c}$ to the applied stress intensity K

Model 1: $$\dot{c} = A_1 \left(\frac{K_1}{K_{1C}}\right)^{n_1} \tag{2.7}$$

Model 2: $$\dot{c} = A_2 \exp\left(n_2 \frac{K_1}{K_{1C}}\right) \tag{2.8}$$

Model 3: $$\dot{c} = A_3 \exp\left(n_3 \frac{K_1^2}{K_{1C}^2}\right) \tag{2.9}$$

Model 4: $$\dot{c} = A_4 \frac{K_1}{K_{1C}} \exp\left(n_4 \frac{K_1}{K_{1C}}\right) \tag{2.10}$$

Model 1 uses the power law crack growth as shown by (2.7). Equations (2.8) to (2.10) show an exponential growth rate in the crack for Models 2 to 4.

The exponential form of the crack growth model (2.8) is based on a physical model that allows the temperature dependence of fatigue to be incorporated in a systematic manner, and this model provides conservative results. Therefore, the crack growth model is recommended over the power law form (2.7). However, for critical uses, a range of models (Models 1 to 4) should be examined to evaluate the range of possible behaviors.

References

[1] Miller, S. E., and A. G. Chynoweth, *Optical Fiber Telecommunications,* New York: Academic Press, 1979.

[2] Kao, C. K., *Optical Fiber Systems,* New York: McGraw-Hill, 1982.

[3] Henry, P. S., "Introduction to Lightwave Transmission," *IEEE Comm. Magazine,* Vol. 23, No. 5, May 1985, pp. 12–16.

[4] Yoshida, S., "Kougaku Gijutsu," *Contact,* Vol. 24, 1986, pp. 681–691.

[5] Baldwin, C. M., R. M. Almeida, and J. D. Mackenzie, "Halide Glasses," *J. Non-Cryst, Solid,* Vol. 43, 1981, p. 309.

[6] Poulain, M., and M. Poulain, *J. Lucas Mater. Res. Bull.,* Vol. 10, 1975, p. 243–343.

[7] Ohsawa, K., T. Shibata, N. Nakamura, and S. Yoshida, *Tech. Digest 7th European Conf. Optical Communication,* Bella Centre, Copenhagen, Denmark, 1981.

[8] Kanamori, T., Y. Terunuma, S. Takahashi, and T. Miyashita, "Chalcogenide Glass Fibers for Mid-Infrared Transmission," *J. Lightwave Tech.,* Vol. LT-2, No. 5, October 1984, pp. 607–613.

[9] Shimizu, M., M. Yamada, T. Takeshia, and M. Horiguchi, *Tech. Digest Optical Amplifiers and their Applications,* Optical Society of America, Washington, DC, Vol. 13, 1991, p. 12.

[10] Izawa, T., S. Kobayashi, S. Sudo, and F. Hanawa, "Continuous Fabrication of High Silica Fiber Preforms," *Int. Conf. Integrated Optics and Optical Communications,* Tokyo, Japan, June 1977.

[11] Inada, Koichi, "Recent Progress in Fiber Fabrication Techniques by Vapor Phase Axial Deposition," *IEEE Trans. Microwave Theory Tech.,* Vol. MTT-30, No. 10, October 1982, pp. 1412–1419.

[12] Sudo, S., M. Kawachi, T. Edahiro, T. Izawa, T. Shioda, and H. Gotoh, "Low-OH-Content Optical Fiber Fabricated by Vapor-Phase-Axial-Deposition Method," *Electron. Lett.,* Vol. 14, 1978, pp. 534–535.

[13] Edahiro, T., M. Kawachi, S. Sudo, and H. Takata, "OH-Ion Reduction in VAD Optical Fibers," *Electron. Lett.,* Vol. 15, August 1979, pp. 482–483.

[14] Chida, K., F. Hanawa, M. Nakahara, and N. Inagaki, "Simultaneous Dehydration with Consolidation for VAD Method," *Electron. Lett.,* Vol. 15, 1979, pp. 835–836.

[15] Edahiro, T., M. Kawachi, S. Sudo, and N. Inagaki, "OH-Ion Reduction in the Optical Fibers Fabricated by the Vapor-Phase Axial Deposition Method," *Trans. IECE Japan,* Vol. E63, August 1980, pp. 8–14.

[16] Inada, K., "Optical Fiber Manufacturing Techniques," *J. IECE Japan,* Vol. 63, 1980, pp. 1150–1156.

[17] Tamura, J., S. Nakamura, E. Kinoshita, A. Lino, and M. Ogai, "Low-Loss Fully Fluorine-Doped Dispersion-Shifted Fibers," *Tech. Digest Conf. Optical Fiber Communications,* Washington, DC, Paper WQ2, 1988.

[18] Ogai, M., and A. Iino, "Low-Loss Dispersion Shifted Fiber with Dual Shape Reflective Index Profile," *Tech. Digest Thirteenth European Conf. Optical Communications,* Helsinki, 1987.

[19] Ogai, M., and A. Iino, "Fully-Fluorine-Doped Single-Mode Fiber and its Application to Submarine Cable," *Tech. Digest Thirteenth European Conf. Optical Communications,* Helsinki, 1987.

[20] Ogai, M, A. Iino, and K. Matsubara, "Characteristics of Fluoride-Doped Silica Fiber," *OFC'88,* January 1988, Paper TUB4.

[21] Sakaguchi, S., M. Nakahara, and Y. Tajima, "Drawing of High-Strength Long-Length Optical Fibers," *Tech. Digest Fourth Int. Conf. Integrated Optics and Optical Fiber Communications,* 1983, pp. 29–45.

[22] Katsuyama, Y., S. Hatano, K. Hogari, T. Matsumoto, and T. Kokubun, "Single-Mode Optical-Fiber Ribbon Cable," *Electron. Lett.,* Vol. 21, No. 4, February 1985, pp. 134–135.

[23] Miyamoto, M., T. Uehara, K. Seto, R. Yamauchi, A. Wada, T. Yamada, O. Fukuda, and K. Inada, "Vapor Axial Deposition Single-Mode Fibers with Ultimate Low-Core Eccentricity," *OFC'87,* January 1987, Paper MC2.

[24] Wada, A., and K. Nishide, *Tech. Digest Thirteenth European Conf. Optical Communications,* Helsinki, 1987.

[25] Wada, A., K. Nishide, M. Horikoshi, R. Yamauchi, and K. Inada, "Two-Step Consolidation for Large Soot Preform," *OFC'88,* January 1988, Paper TUB2.

[26] Sudo, S., T. Edahiro, and M. Kawachi, "Sintering Process of Porous Preforms Made by a VAD Method for Optical Fiber Fabrication," *Trans. IECE Japan,* Vol. E83, No. 10, 1980, pp. 731–737.

[27] Japanese Delegation, Presentation on Advances on Satellite Communications in Japan, Industry Canada, St. Hubert, 1993.

[28] Anesaki, N., *Aerospace Industry in Japan,* The Society of Japanese Aerospace Companies, Inc., Tokyo, Japan, 1993–1994.

[29] Blankenship, M. G., A. J. Morrow, and L. A. Silverman, "Large Graded Index Preforms Deposited at High Rate Using Outside Vapor Deposition," Presented at OFC'82, Phoenix, AZ, April 13, 1982.

[30] Schultz, P. C., "Fabrication of Optical Waveguides by Outside Vapor Deposition Process," *Proc. IEEE,* Vol. 68, October 1980, pp. 1187–1190.

[31] Blankenship, M. G., U.S. Patent 4,251,251, 1981.

[32] Bailey, A. C., and A. J. Morrow, U.S. Patent 4,298, 365, November 1981.

[33] Blankenship, M. G., and C. W. Deneka, "The Outside Vapor Deposition Method of Fabricating Optical Waveguide Fibers," *IEEE Trans. Microwave Theory Techniques,* Vol. MTT-30, No. 10, October 1982, pp. 1406–1411.

[34] Berkey, G. E., "Fluorine-Doped Fibers by the Outside Vapor Deposition Process," *OFC'84,* January 1984, Paper MG3.

[35] Keck, D. B., P. C. Shultz, and F. Zimar, U.S. Patent 3373292 1973.

[36] MacChesney, J. B., and P. B. O'Connor, U.S. Patent 4,217,072 1980.

[37] Wood, D. L., K. L. Walker, J. R. Simpson, J. B. MacChesney, D. L. Nash, and P. Angueira, *Tech. Digest Seventh ECOC,* Copenhagen, Denmark, 1981, pp. 1.2.1–1.2.4.

[38] Nagel, S. R., J. B. MacChesney, and K. L. Walker, "An Overview of the Modified Chemical Vapor Deposition (MCVD) Process and Performance," *IEEE Trans. Microwave Theory Tech.,* Vol. MTT-30, No. 4, April 1992, pp. 305–322.

[39] Nagel, S. R., and J. W. Fleming, "Latest Developments in Fiber Manufacture by MCVD and RFMCVD," *OFC'84,* January 1984, Paper TUI1, p. 52.

[40] McAffe, K. B., K. L. Walker, R. A. Laudise, and R. S. Hozack, "Modified Chemical Vapor Deposition Process Chemistry," *OFC'84,* January 1984, Paper TUI2, pp. 52–54.

[41] MacChesney, J. B., D. W. Johnson, Jr., P. J. Lemaire, L. G. Cohen, and E. M. Rabinovich, "Fluorosilicate Substrate Tubes to Eliminate Leaky-Mode Losses in MCVD Single-Mode Fibers with Depressed-Index Cladding," *Tech. Digest Conf. Optical Fiber Communications,* Washington, DC, 1985, Paper WH2.

[42] Walker, K. L., "Current Status of the MCVD Process," *OFC'86,* February 1986, Paper WA3, p. 78.

[43] Poole, S. B., D. N. Payne, R. J. Mears, M. E. Fermann, and R. I. Laming, "Fabrication and Characterization of Low-Loss Fibers Containing Rare-Earth Ions," *IEEE/OSA J. Lightwave Tech.,* Vol. LT-870, 1986, pp. 870–880.

[44] Walker, K. L., "Current Status and Future Trends in Optical Fiber Fabrication," *OFC'91,* February 1991, Paper WA1, p. 61.

[45] Gardner, W. B., A. A. Klein, and H. T. Shang, "Low Polarization Dispersion in MCVD Dispersion-Shifted Fibers," *OFC'91,* February 1991, Paper WA5, p. 65.

[46] Gardner, W. B., A. A. Klein, and T. Shang, "Low Polarization Dispersion in MCVD Dispersion-Shifted Fibers," *OFC'91,* February 1991, Paper WA5, p. 65.

[47] Fleming, J. W., "Sol-Gel Techniques for Lightwave Applications," *Tech. Digest Sixth Int. Conf. Integrated Optics and Optical Fiber Communications,* Washington, DC, 1985, Paper MH1.

[48] MacChesney, J. B., "Optical Fiber Fabrication by the Sol-Gel Method," *OFC'88,* January 1988, Paper TUB1.

[49] DiGiovanni, D. J., and J. B. MacChesney, "New Optical Fiber Fabrication Technique Using Sol-Gel Dipcoating," *OFC'91,* February 1991, Paper WA2, p. 62.

[50] Morse, T. F., A. Kilian, and W. Risen, Jr., "Aerosol Techniques for Fiber Core Doping," *OFC'91,* February 1991, Paper WA3, p. 63.

[51] Jaeger, R. E., J. B. MacChesney, and T. J. Miller, "The Preparation of Optical Waveguide Preforms by Plasma Deposition," *Bell Syst. Tech. J.,* Vol. 57, No. 1, 1978, pp. 205–210.

[52] Irven, J., and A. Robinson, "Optical Fibers Prepared by Plasma Augmented Vapor Deposition," *Phys. Chem. Glasses,* Vol. 21, No. 1, 1980, pp. 47–52.

[53] Fleming, J. W., and V. R. Raju, "Low Optical Attenuation Fiber Prepared by Enhanced MCVD," *Tech. Digest European Conference on Optical Communications,* San Francisco, CA, 1981, Paper WD2.

[54] Kats, A., "Glass-Outline of a Development," *Philips Tech. Rev.,* Vol. 42, No. 10/11/12, September 1986, pp. 316–324.

[55] Nagel, S. R., "Progress in the PCVD Process," *OFC'86,* February 1986, Paper WA1, p. 76.

[56] Lydtin, H., "PCVD: A Technique Suitable for Large-Scale Fabrication of Optical Fibers," *J. Lightwave Tech.,* Vol. LT-4, 1986, p. 1034.

[57] Griffioen, W., "Evaluation of Optical Fiber Lifetime Models Based on the Power Law," *Optical Engineering*, Vol. 38, No. 2, 1994, pp. 488–497.

[58] Bauch, H., V. Paquet, and W. Siefert, "Influence of the Preform Collapse Process on the Attenuation of PICVD fibers with Pure SiO_2 Core," *OFC'88,* January 1988, Paper TUB3.

[59] Nagasawa, K., H. Yutaka, Y. Ohki, and K. Yahagi, "Improvement of Radiation Resistance of Pure Silica Core Fibers by Hydrogen Treatment," *Japan J. Appl. Phys.*, Vol. 24, 1985, pp. 1224–1228.

[60] Wei, T., M. P. Singh, W. J. Miniscalco, and J. A. Wall, "Effect of Hydrogen Treatment on Radiation Hardness of Optical Fibers," *Proc. SPIE*, Vol. 721, 1987, pp. 32–36.

[61] Evans, B. D., "The Role of Hydrogen as a Radiation Protection Agent at Low Temperature in a Low-OH, Pure Silica Optical Fiber," *IEEE Trans. Nuclear Sci.*, Vol. NS-35, 1988, pp. 1215–1220.

[62] Oe, M., M. Wanatabe, G. Tanaka, K. Tsuneshisa, Y. Chigusa, I. Yoshiki, and Y. Ishiguro, Japanese Patent 0,1279,207, 1989.

[63] Huff, R. G., and F. V. DiMarcello, "Hermetically Coated Optical Fiber for Adverse Environments," *Proc. SPIE*, Vol. 867, 1987, pp. 40–45.

[64] Miller, A. E., M. F. Yan, H. A. Watson, and K. T. Nelson, "Radiation-Hardened Optical Fibers for Low Temperature High-Dosage Environments," *OFC'92*, 1992, Paper ThF4, p. 216.

[65] Nagel, S. R., "Reliability Issues in Optical Fibers," *Proc. SPIE*, Vol. 717, 1986, pp. 8–20.

[66] Kurkjian, C. R., J. T. Krause, and M. J. Matthewson, "Strength and Fatigue of Silica Optical Fibers," *J. Lightwave Tech.*, Vol. LT-7, No. 9, September 1989, pp. 1360–1370.

[67] Krause, J. T., et al., "Mechanical Reliability of Hermetic Carbon Coated Optical Fiber," *EFOC/LAN-88 Proc.*, 1988, p. 121.

[68] Zhurkov, S. N., "Kinetic Concept of the Strength of Solids," *Int. J. Fract. Mech.*, Vol. 1, 1965, pp. 311–322.

[69] Bubnov, M. M., E. M. Dianov, and S. L. Semjonov, "Maximum Value of Fatigue Parameter *n* for Hermetically Coated Silica Glass Fibers," *OFC'92*, 1992, Paper ThF2, p. 216.

[70] Bubel, G. M., and M. J. Matthewson, "Optical Fiber Reliability Implications of Uncertainty in the Fatigue Crack Growth Model," *Opt. Eng.*, Vol. 30, No. 6, 1991, pp. 737–745.

[71] Matthewson, M. J., "Estimating Mechanical Reliability Optical Fiber," *OFC'92*, 1992, Paper ThF1, p. 215.

Chapter 3

Radio Frequency Beamforming/Scanning Techniques

3.1 INTRODUCTION

This chapter presents a brief description of various technologies used for RF/microwave beamforming/scanning. The beamformer/scanner types covered include the dielectric (including nonuniform refractive index) and waveguide lenses, the Ruze lens, the bootlace concept, and the Rotman, R-2R, R-KR, objective lenses. Also covered are circuit beamformers. Circuit types of beamformer/scanner presented include the Blass and the Butler matrices. The single and the dual-reflector imaging antennas with electronic scanning are described. A short description of the true-time delay (TTD) and the injection-phase-locking concepts are discussed. These concepts are used for beamforming/steering of *phased array antennas* (PAAs) in Chapters 5 and 6.

3.2 LENS-BASED BEAMFORMING/SCANNING TECHNIQUES

In recent years, lens-based beamforming/scanning technology has become very promising and useful for radar, ground-based, and space-based communications antennas. The design of such antennas is seriously complicated by the requirements of controlling not only in phase but also the amplitude distribution over the array while scanning the antenna beam. A brief summary of lens type of antennas is given below.

3.2.1 Dielectric Lenses

Dielectric lenses can be used for microwave antennas in place of the more common parabolic reflector because both are fundamentally equivalent; that is, they both convert a spherical phase front from the feed to a plane phase-front from the aperture. The main advantage of lens antennas is that the feed does not cause aperture blocking. Given the feed point location and illumination, reflector radiation pattern control is limited to choosing the reflector contour, but even lenses with constant dielectric properties have two variable contours, that is, the front and back lens surfaces. However, simple optical dielectric lenses do not have very wide angle capability. For wide-angle scanning, a nonuniform refractive index low-loss dielectric material is introduced between the focus and objective aperture of a lens system. This type of lens can provide much wider scan multiples than reflectors because it inherently offers more degrees of freedom in design. For wide-angle scanning or wide-angle multiple beams, the spherical Luneburg lens is useful. This lens is not made of a uniform (homogeneous) dielectric like the simpler optical lenses but has a nonuniform refractive index, varying from the center to the edge according to the ''Luneburg law'' [1]. Figure 3.1 shows a Luneburg lens that formed a focus at infinity, and the variation of refractive index n with radius r can be given by

$$n(r) = [2 - r^2]^{1/2} \tag{3.1}$$

where n varies from $(2)^{1/2}$ at the center to 1 at the lens edge. For 360° scanning the feed can be moved around the $r = 1$ contour. Gutman [2] showed that less index variation is required in the central region if an outer shell of constant refractive index is used. Morgan [3] simplified the construction of the lens by making a fully layered structure. Such lenses are commercially available and can provide two-dimensional scanning. Cassel [4] reported a one-dimensional microwave broadband antenna with 360° multiple-beam coverage, and Vogel [5] reported a planar waveguide Luneburg lens for the millimeter-wave region. The equivalent geodesic realization of such lenses is achieved using parallel plate regions with nonplanar contours. DuFort [6] reported a wide-angle scanning antenna that

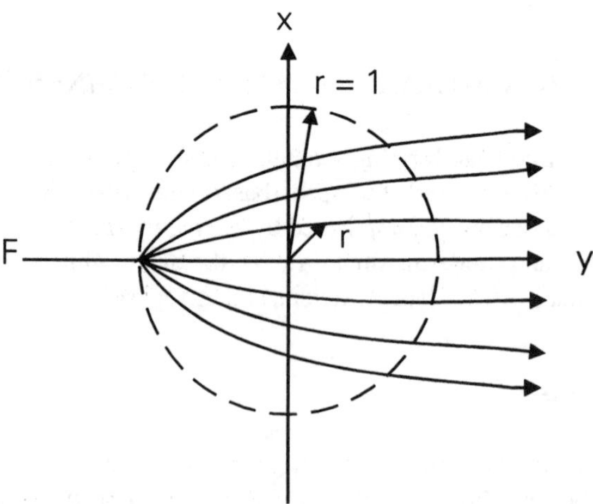

Figure 3.1 Luneburg lens.

provides a 360° scan of a 3°-wide beam with −15-dB sidelobes at 26 GHz, using a Rinehart lens. The Rinehart lens is the geodesic analog of the Luneburg lens. Figure 3.2 shows the profile of the Rinehart lens. Its profile length, *s*, as a function of radial distance, *r*, takes the simple form

$$s = (r + \arcsin(r))/2 \qquad 0 \leqslant r \leqslant 1 \tag{3.2}$$

However, this transforms into the following rather formidable expression in cylindrical coordinates:

$$z = (1/2)\sqrt{(4 - 3r^2 + 4f(r))} - (1/\sqrt{3}) \ln \sqrt{(3/2 + (3/2)f(r))} \tag{3.3}$$

where *s* and *z* are shown in Figure 3.2 and

$$f(r) = [(1 - r^2)]^{1/2} \qquad 0 \leqslant r \leqslant 1$$

The height of the lens is 31.6% of its diameter, and its entry/exit annulus faces vertically downward. This opens up the possibility of feeding the emergent energy into a cylindrical parallel plate structure with radiating slots in its outer face or, alternatively, into a cylindrical array of slotted waveguide radiators. Wild [7] showed in his patent that radiators could be curved in vertical profile to achieve a vertical distribution of radiated power suitable for the *microwave landing system* (MLS). This type of configuration can scan a beam over a sector of ±60° without diplexing the lens connections or could achieve 360° scanning if a full circle of radiators were provided and circulators or diplexing switches were introduced into the interface between the lens and the radiating elements.

3.2.2 Waveguide Lenses

This type of lens, as its name implies, is composed of waveguide elements whose physical lengths are determined by the condition that the electrical path length of a general ray from a source at the focus to a plane perpendicular to the axis of the lens is equal to the path

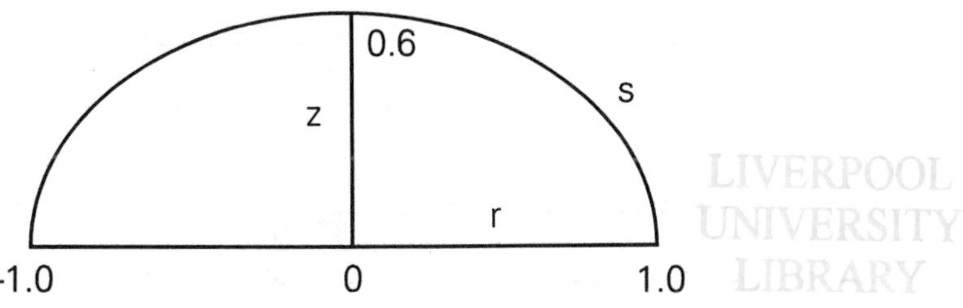

Figure 3.2 Profile of the Rinehart lens.

length of the central ray. Important applications of waveguide lens antennas are for beamforming/scanning applications. This concept was not feasible some years ago, but with several new light-weight materials (for example, titanium) and their derivatives, it is possible to design multibeam communication antennas in the frequencies of Ku-band and higher.

Figure 3.3 shows a typical waveguide lens. Instead of the shape of an original lens the lens is thicker at the edges than the center, because optical lenses delay the ray paths while waveguide lenses speed up the ray paths since the phase velocity in a waveguide is faster than *C*, the speed of light.

In microwave antennas, where size and weight are limiting factors, the thickness of a waveguide lens is reduced by zoning the contour of the lens. The zoning reduces the length *d* (Figure 3.4) by a factor of multiple *m* integral guide wavelengths ($m = 1, 2, 3, \ldots$) without changing the electrical performance of the lens. In fact, zoning increases the bandwidth about three times more than an unzoned lens. The minimum thickness waveguide lens greatly reduces the thickness and, hence, the weight of the lens.

In 1950, Ruze [8] proposed a wide-angle metal-plate lens. Ruze showed that the rays within the lens travel parallel to the axis, that is, the path P–Q as shown in Figure 3.5. The path P–Q can be formed in a variety of ways such as waveguide, coaxial line, triplate stripline, or microstrip. The electrical path length P–Q is varied across the lens dimension, *y*. Moving point C radially toward or away from the lens, which is known as refocusing, reduces the phase error. Increasing the focal length also reduces aberration.

Figure 3.6 shows four types of Ruze lenses. The constant thickness lens, Figure 3.6(c), is mechanically compact but has an excessively curved focal arc, whereas for that in Figure 3.6(b) no focusing is required. The constant refractive index form of Figure 3.6(d) results in a highly curved focal arc and a thick lens that can be thinned by zoning although this results in step phase errors. Figure 3.6(a) shows the flat-face lens for which

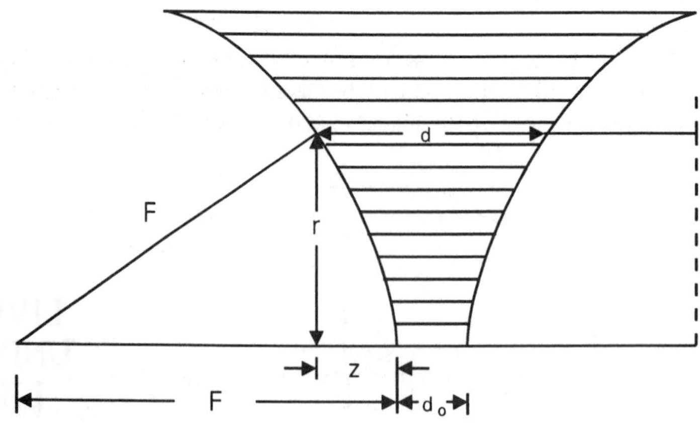

Figure 3.3 Sketch of a typical waveguide lens.

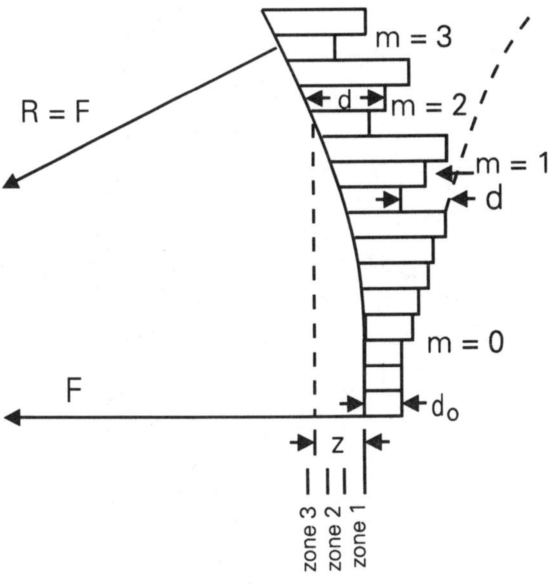

Figure 3.4 Waveguide lens zoned for minimum thickness.

small refocusing results in very low aberrations. A general rule in designing these lenses with more than one perfect focus is: for each focal point there is a path length equality condition. The scanning mechanism of the lens is based on equal path lengths; the scan direction is thus frequency insensitive. Multifocal points provide the lens with its wide-angle scanning properties.

3.2.3 The Bootlace Lens Concept and the Rotman Lens

This type of lens consists of *transmission lines* (TEM mode) that interconnect a pick-up array of small radiators on the lens primary, inner surface with a similar array of radiators on the secondary, or outer, contour of the lens. This configuration is fundamentally nondispersive (no cut-off wavelength of waveguide elements). Fundamental limitations are the impedance mismatch between elements and the bandwidth of radiating elements. This lens is capable of 30% to 60% bandwidth. Common realizations of the TEM line are round coaxial cable and a lens consisting of literally thousands of such cables interconnecting small dipoles or horn radiators on inner and outer surfaces as shown in Figure 3.5(a). The fundamental difference between the Ruze lens and the bootlace lens is that the electrical length P–Q′ is no longer parallel to the axis as shown in Figure 3.5(b). Here the three variable parameters of the flat-face Ruze design—namely lens contour, thickness, and refractive index—are supplemented by a fourth, that is, the relative front and back

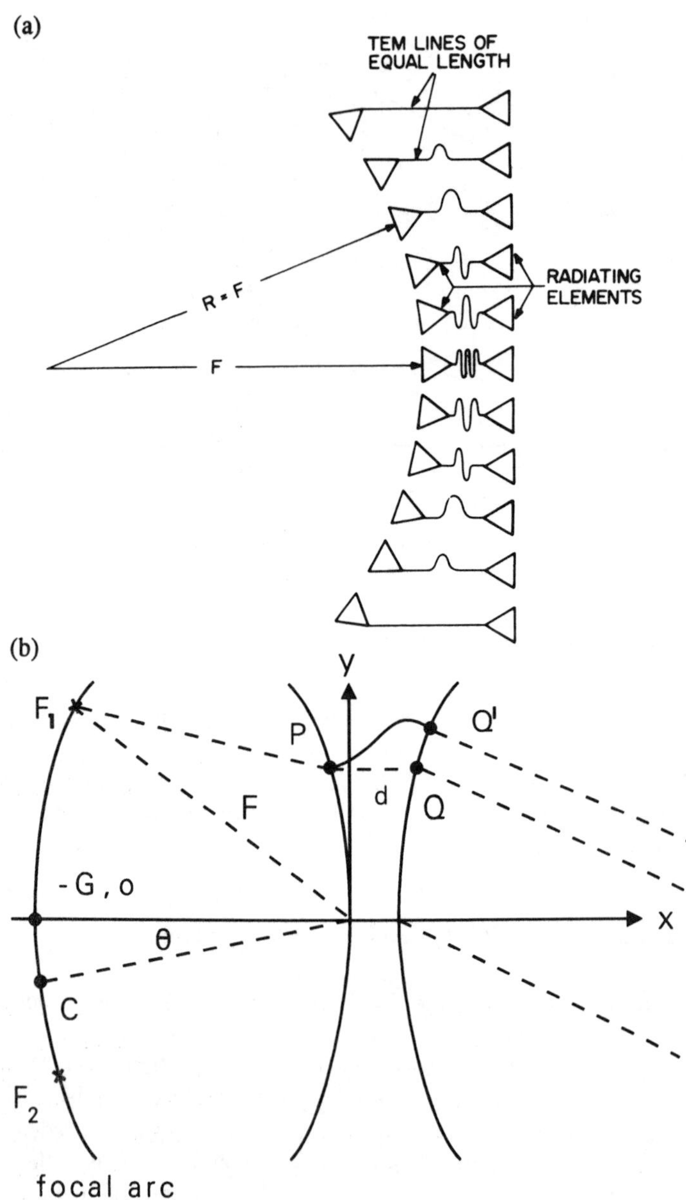

Figure 3.5 (a) Bootlace lens concept. (b) Geometry of the lens: P–Q specifies Ruze form and P–Q′ specifies Rotman type. (© 1982 IEE.)

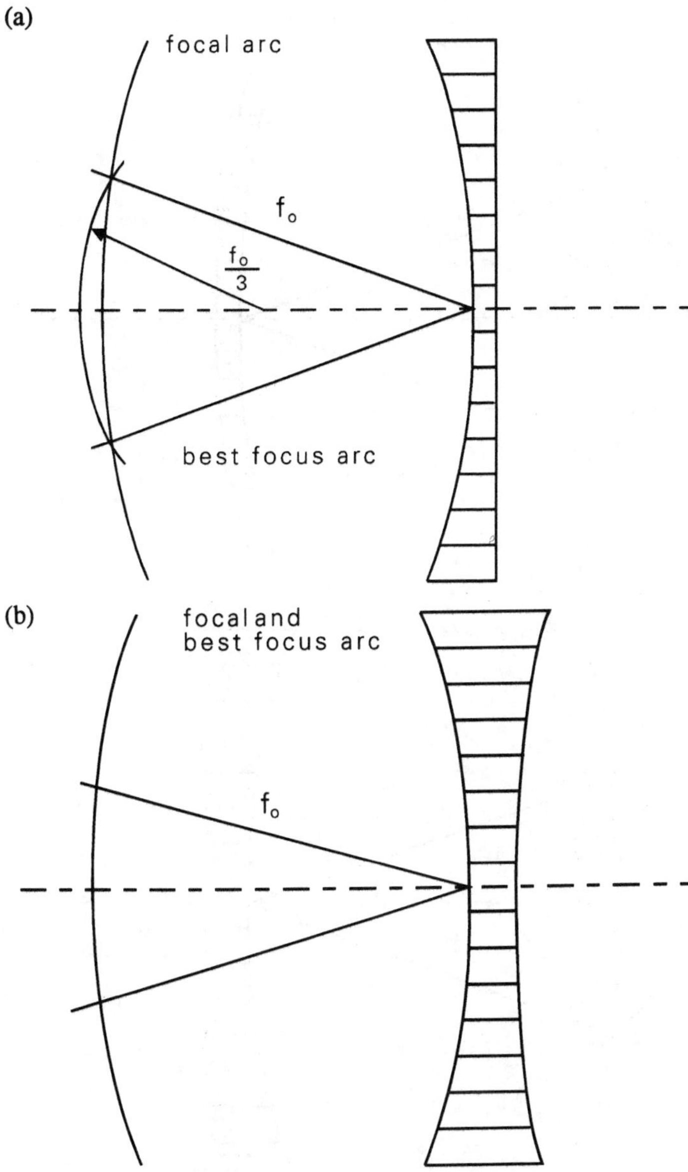

Figure 3.6 Ruze lens: (a) straight front face; (b) no second order (triple correction); (c) constant thickness; (d) constant refractive index.

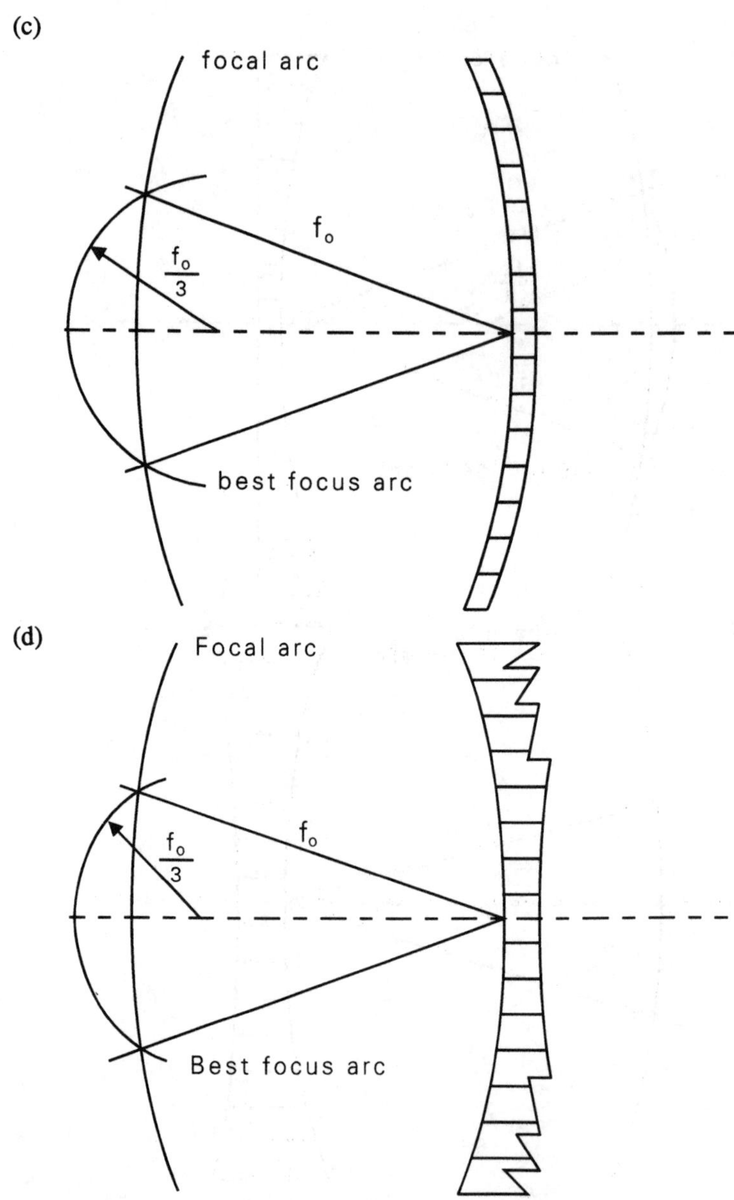

Figure 3.6 (continued)

element spacing as described by Rotman et al. [9] and Niazi [10]. For a particular scan requirement the angle subtended by F_1 and F_2 and the normalized focal length (G/F) must be specified. The angle α is usually taken as the maximum required scan because the phase error is minimized between $\pm\alpha$ but increases significantly beyond it. The optimum G for minimum aberration is given by Smith [11] as

$$F/G = \cos\alpha \tag{3.4}$$

But the total aberrations for the two cases ($G/F = 1$ and 1.137) are shown in Figure 3.7(a, b) [12], which are similar and of the order of 6×10^{-4} G. The contour shape of the lens controls the amplitude and multiple reflections.

Rotman et al. [9], Jasik [13], Archer [14], Kumar [15], and Smith [16] have used Rotman lenses to form multiple beams for antenna arrays. They employ path-length-difference beam deflection, so that the beam direction remains stationary with frequency in contrast to a Butler matrix [17]. A Rotman lens feeding a linear array is shown in Figure 3.8, which uses TEM propagation between flat parallel plates. A variety of curves can be adopted for the input and output arcs, giving a trifocal design [9] with perfect collimation of the radiated beam for an on-axis and two symmetrically placed off-axis feed points. The lens dimensions then control the amount of phase error at other beam angles and, for acceptable performance over the scan angles typical of interscan and the transverse fore and aft dimensions of the lens, become compatable with the array width.

Figure 3.9 shows the multiple-beam radiation pattern at 10 GHz predicted from the measured amplitudes and phases at the array ports of the experimental lens. The lens provides five beams (the beams at −30°, −15°, and 0° are shown in Figure 3.9; the beams at +15° and +30° are related by symmetry), with sidelobe levels equal to −20 dB and an insertion loss equal to 2 dB, over the 8-GHz to 12-GHz frequency band. Since 1982, many researchers [19–30] reported various improvements on Rotman lenses such as low sidelobes of −29 dB and −33 dB in single beams and sidelobe levels of −16 dB with a −5-dB amplitude taper across the array arc for scanning out to + 45° in X-band.

Sole et al. [31] demonstrated a two-dimensional configuration of Rotman lenses that can be used to feed a planar array. Figure 3.10 shows that one stack of the lens performs elevation beamforming, with the second stack performing azimuth beamforming. Alternatively, a single volumetric or three-dimensional Rotman lens is discussed by many researchers [32–34]. Sole et al. [31] used 12 and 37 horns on the beam and array port surfaces, respectively, and they measured sidelobes of less than −20 dB out to scan angles of ±30°.

3.2.4 Lense Beamformer for Conformal Arrays

If the flat front-face condition is removed from the Rotman lens, then the following lenses for conformal array result [35]: (1) R-2R lens and (2) R-KR lens.

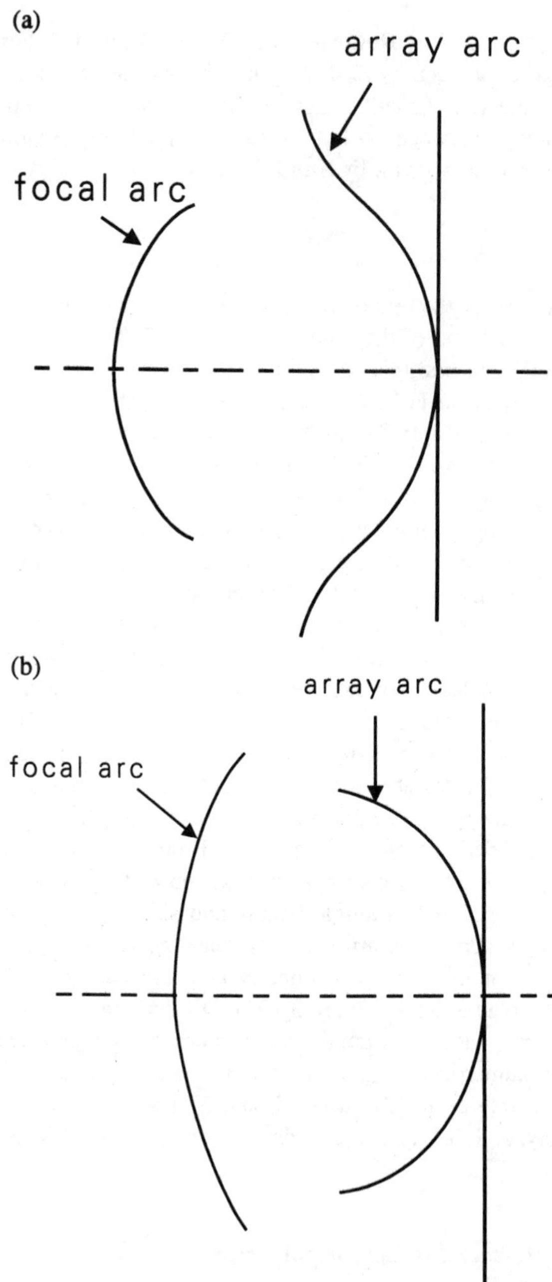

Figure 3.7 Rotman lens contour for α = 30° and (a) G/F = 1.0 and (b) G/F = 1.137.

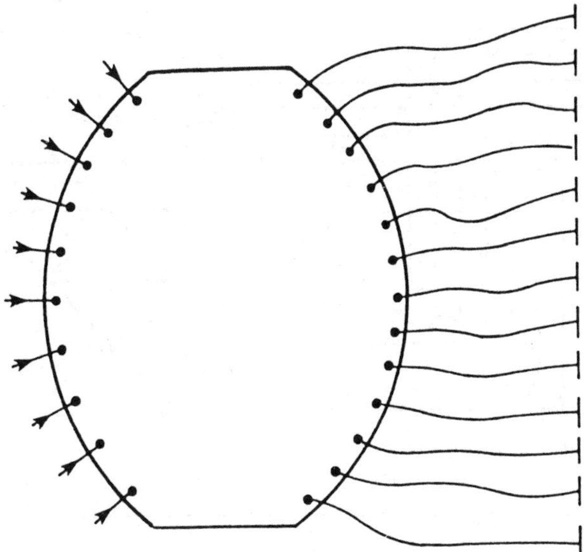

Figure 3.8 Schematic diagram of a Rotman lens fed linear array.

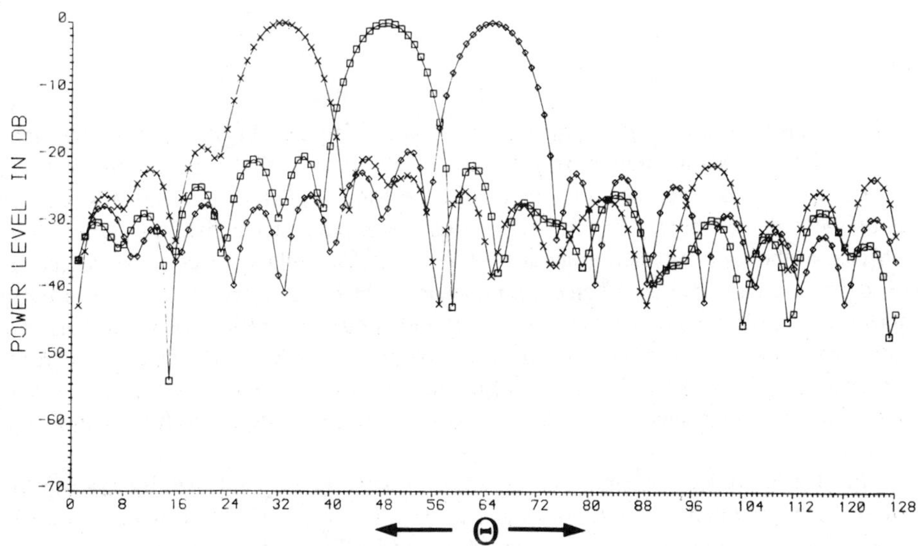

Figure 3.9 Multiple-beam radiation patterns derived from the experimental amplitude and phase distributions at 10 GHz for the waveguide-fed Rotman lens. (© 1983 IEE.)

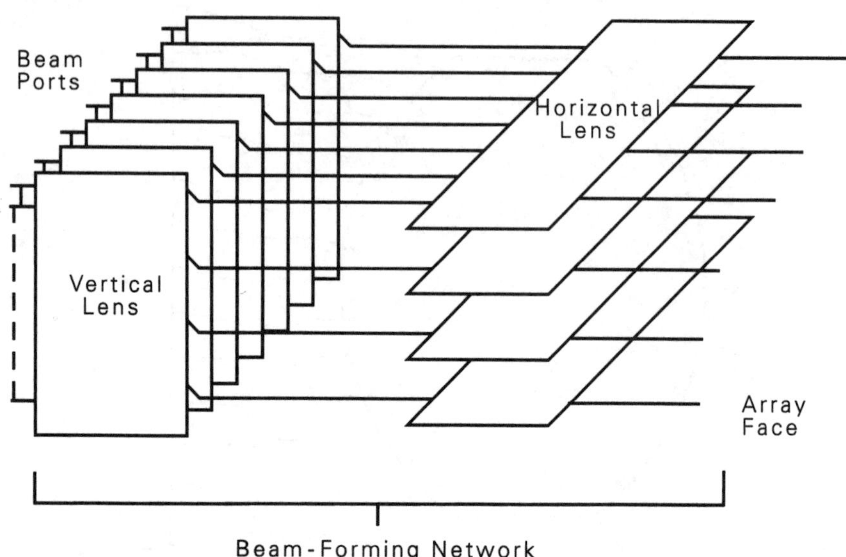

Figure 3.10 Two-dimensional Rotman lens feeding a planar array. (© 1987 IEE.)

R-2R Lens

The R-2R lens [36] has circular symmetry as shown in Figure 3.11. It contains a disc lens of radius *R* that feeds a circular array of radius 2*R* through equal-length cables. Perfect collimation of the radiated beam is obtained if the angular pitch of the lens taps is twice that of the array elements. In this case, a lens feed point, located at an angle, Θ, is connected to the array element at an angle Θ/2. Therefore, when the point on the lens is moved by an angle, Θ, the radiated beam scans by Θ/2. Equal-length lines are used to connect a pick-off point *n* on the lens to the radiating element *n*. The R-2R lens is a limiting case of a family of lenses shaped as spherical caps in which the angular transformation ratio, *K*, between the array and the lens is progressively varied from 0.5 to 1, where $K = 0.5$ corresponds to the R-2R lens with infinite radius of curvature as shown in Figure 3.11.

For the K-2R lens, scanning of the antenna beam is accomplished by moving the feed along the perimeter of the lens surface [37]. The beam direction is diametrically opposite the feed point. Archer [14] showed that the radiating elements extending out to +90° on the array must be excited by pick-off points extending out to +180° on the lens. The scanning of the feed and, hence, the antenna beam, is not possible because pick-off points occupy all of the lens' perimeter space and no place is left for feed connection points. Therefore, the lens cannot illuminate more than half the arc of a circle.

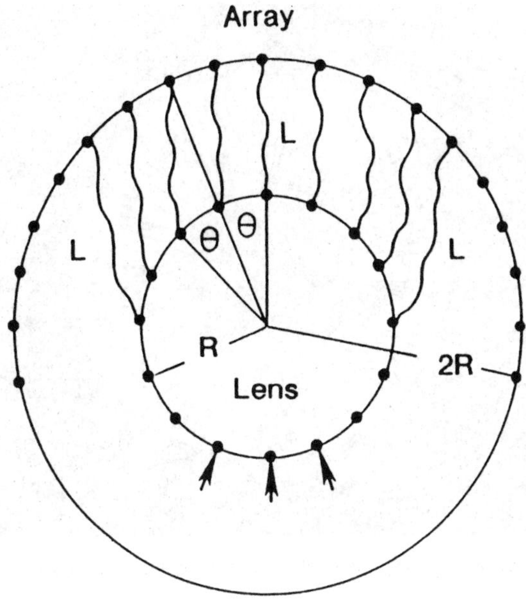

Figure 3.11 Geometry of the R-2R lens.

R-KR Lens

The R-KR lens provides a full 360° of beam coverage by feeding a circular array. Archer [14] showed a 100-element array-fed lens that operates over the 4-GHz to 11-GHz frequency band. Figure 3.12 shows the measured radiation patterns of a 100-element array fed by a R-KR lens at 11 GHz. The array ports of the lens are one and the same; the full connection to the circular array does not provide any output ports. A three-port circulator in each of the connecting lines between the lens array may be used to create a set of output beam ports. This method does not provide wide bandwidth. Therefore, for cases where wide-bandwidth circular arrays are required, the approach shown in Figure 3.13 is recommended. Bodnar et al. [38–40] used a 90° hybrid at L-band, as shown in Figures 3.13 and 3.14. Figure 3.13 shows that a transmitting signal input to port 24 has its power split evenly between the upper A_1 and lower A_2 lenses by a 90° hybrid with the input to the upper lens advanced 90° in phase with respect to that of the lower lens. These lenses distribute the split signal to feed points around the periphery, where the signals are recombined by 90° hybrids. The signal from the upper lens adds in phase to a signal advanced by 90° from the lower lens to provide an output sum signal to the radiating elements on the element layer. At the hybrids, these signals add out of phase so that no power goes to port 18 and element 24 is isolated from the input signal at beam port 24. The circuit operates in a similar way for the receiving mode.

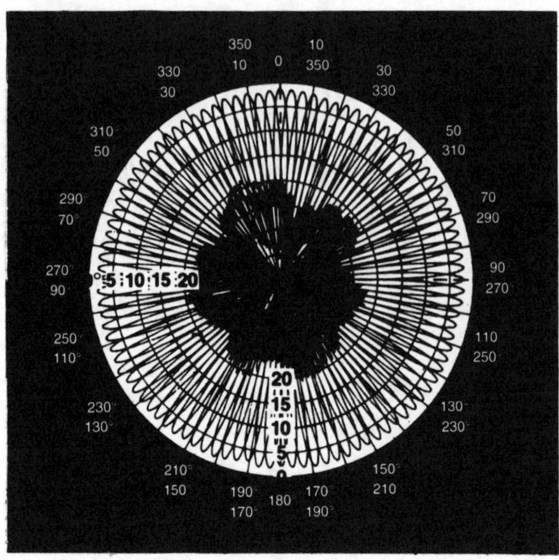

Figure 3.12 Measured radiation patterns of a 100-element array fed by an R-KR lens at 11 GHz [35].

Figure 3.14 shows geometric relationships for the R-KR lens. It has a lens radius *KR,* and the radiating elements on the aperture are located on a circle of radius, *R.* In the figure, both the radiating element and the lens pick-off points are located at the same angle, α. The total electric path from the feed point to the desired wavefront is given by

$$2\pi(L/\lambda_G + d/\lambda + t/\lambda_t) \tag{3.5}$$

where λ_G, λ, and λ_t are the wavelengths in the lens, air, and transmission line, respectively, and t is the length of transmission line from the pick-off points to the radiating element. The parameters d, L, and K are given in [35] as

$$d = R(1 - \cos\alpha) \tag{3.6}$$

$$L = 2KR\cos(\alpha/2) \tag{3.7}$$

$$K = 1.9 \quad \text{for } k_G L + kd \approx \text{constant (when } k_G = k) \tag{3.8}$$

where $K_G = 2\pi/\lambda_G$ and $K = 2\pi/\lambda$.

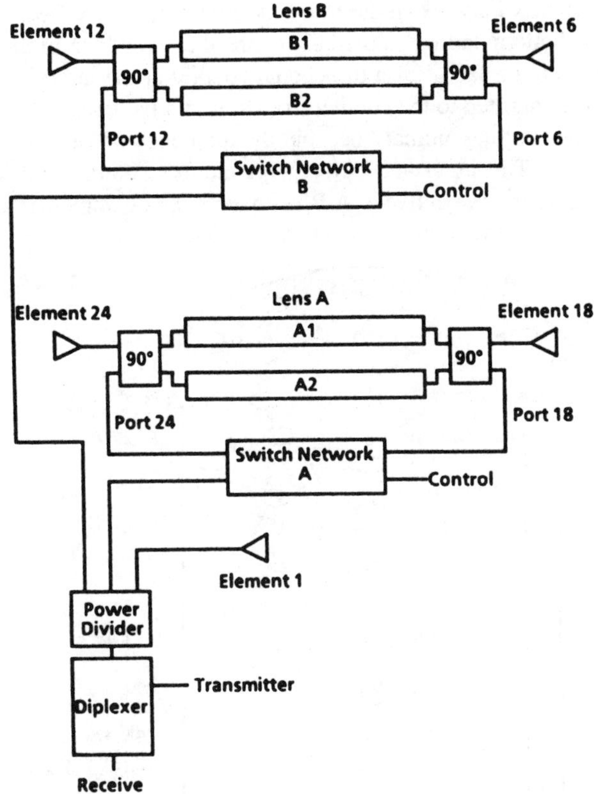

Figure 3.13 Feed network for an R-KR array antenna [35].

3.3 CIRCUIT BEAMFORMERS

Beam scanning of the antenna array requires circuit beamformers, which use couplers, power splitters, and transmission lines. In this system, the phase shift is produced by the lengths of the transmission lines and aperture amplitude distributions are controlled by the power splitter ratios. There are two types of beamformers: (1) the Blass matrix and (2) the Butler matrix.

3.3.1 The Blass Matrix

The Blass matrix consists of a number of traveling wave feed lines connected to a linear array through another set of lines as described by Blass [41]. The required number of amplifiers need not be binary but may be reduced to equal the number of active ports. On the other hand, it requires a coupler at every cross point. Figure 3.15 shows a simple sketch of a Blass matrix network in the transmission mode. In the figure the longitudinal lines are fed at the bottom and RF is partially coupled to every one of the transverse lines according to the coupling factor of a directional coupler at the intersection. The electromagnetic waves are admitted to the coupler from longitudinal lines to transverse lines and from transverse lines to longitudinal lines, but the directions of the flow are only upward or from left to right. The input and output port numbers correspond to the beam and antenna element numbers, respectively. A Blass matrix can be implemented in waveguide,

(a)

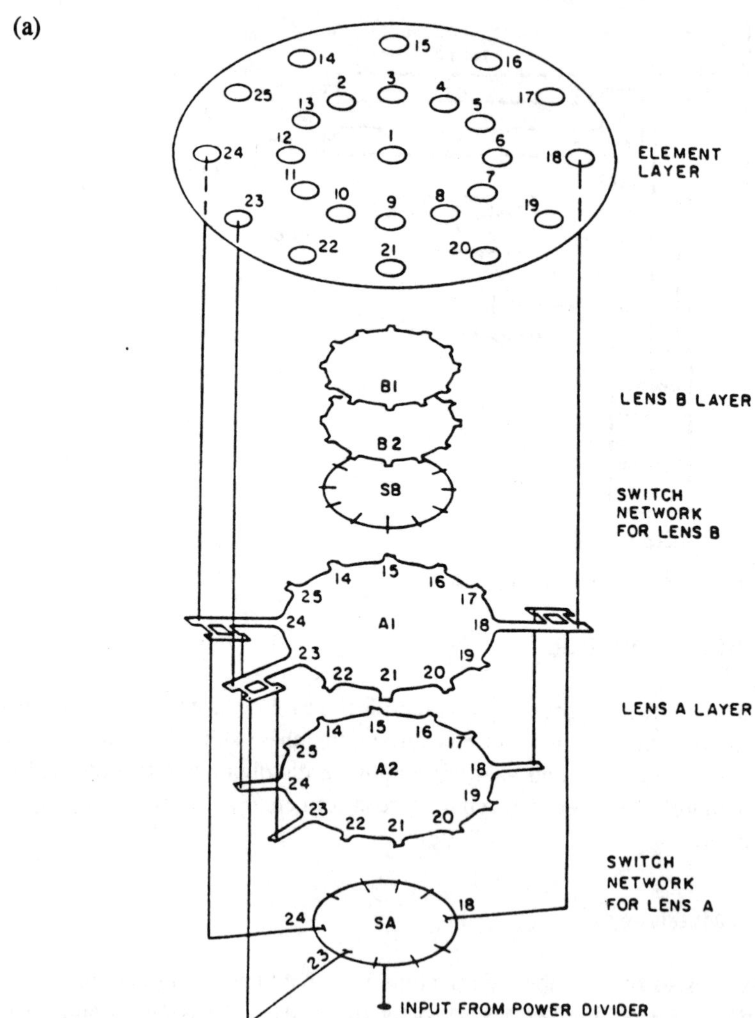

Figure 3.14 Geometry of the (a) R-KR lens-fed array antenna and (b) the R-KR lens [35].

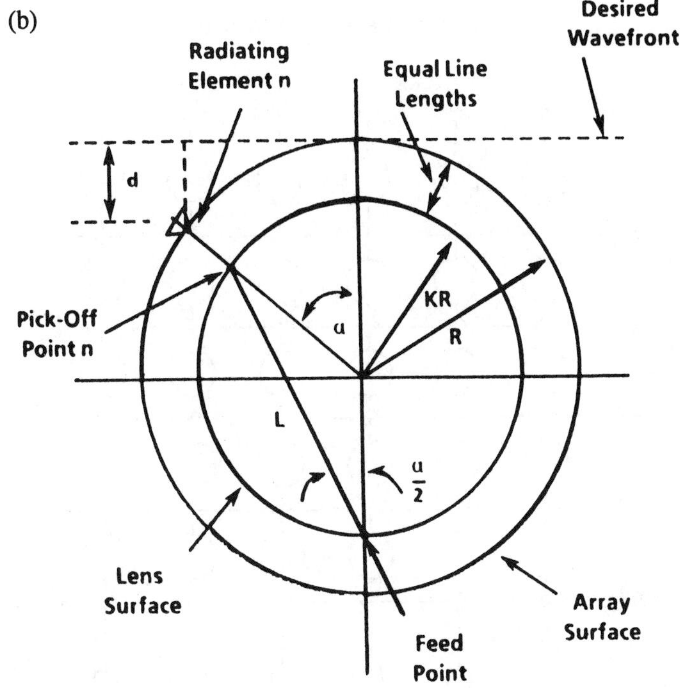

Figure 3.14 (continued)

strip line, or TEM-line technology. A signal applied at a beam port travels along the feed line to the end termination. At each crossover point a small signal is coupled into each element line, which excites the corresponding radiating element. The path difference between the input and each element controls the radiated beam direction. Due to the traveling wave nature of the network, the input match will be good and the beam set will scan with frequency. The important parameters in designing the network consist of selecting the appropriate phase and coupling values to achieve the desired beam set.

The Blass matrix concept has been extended to form a planar, two-dimensional multiple-beam microstrip patch array [42]. A typical scan range of ±60° has been achieved for an aperture size of 15 wavelengths.

3.3.2 The Butler Matrix

Butler et al. [17] reported the Butler matrix circuit beamformer consisting of interconnected fixed phase shift section and 3-dB hybrid couplers as shown in Figure 3.16. The dashed lines show that a signal into one port divides to excite the radiating elements, with equal amplitude and linear phase taper. The Butler matrix network provides a good input

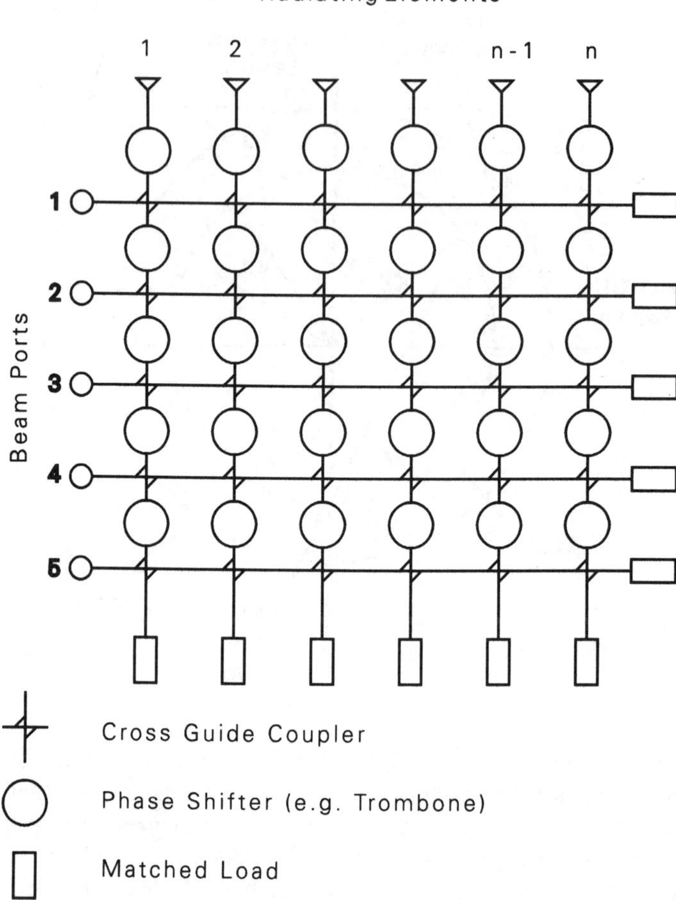

Figure 3.15 Sketch of a Blass matrix.

VSWR and beam isolation, Shelton et al. [43] and Muenzer et al. [44] improved the flexibility of the Butler matrix by use of six- and eight-port hybrid couplers and termination of unused ports. In 1978, Macnamara [45] introduced 180° hybrids to reduce the number of phase shifters significantly; for example when $N = 32$, 15 are saved. Shelton et al. [43] and Guy et al. [46] reported a further reduction in the hybrid count by using reflective matrices. In 1984, Blokhia et al. [47] removed the crossovers (Figure 3.16) by reconfiguring into a checker-board network. Many researchers [48–51] constructed the Butler matrix network in various media, including waveguides for high-power use and microstrip. The matrix has also been used as a commutating device in circular and

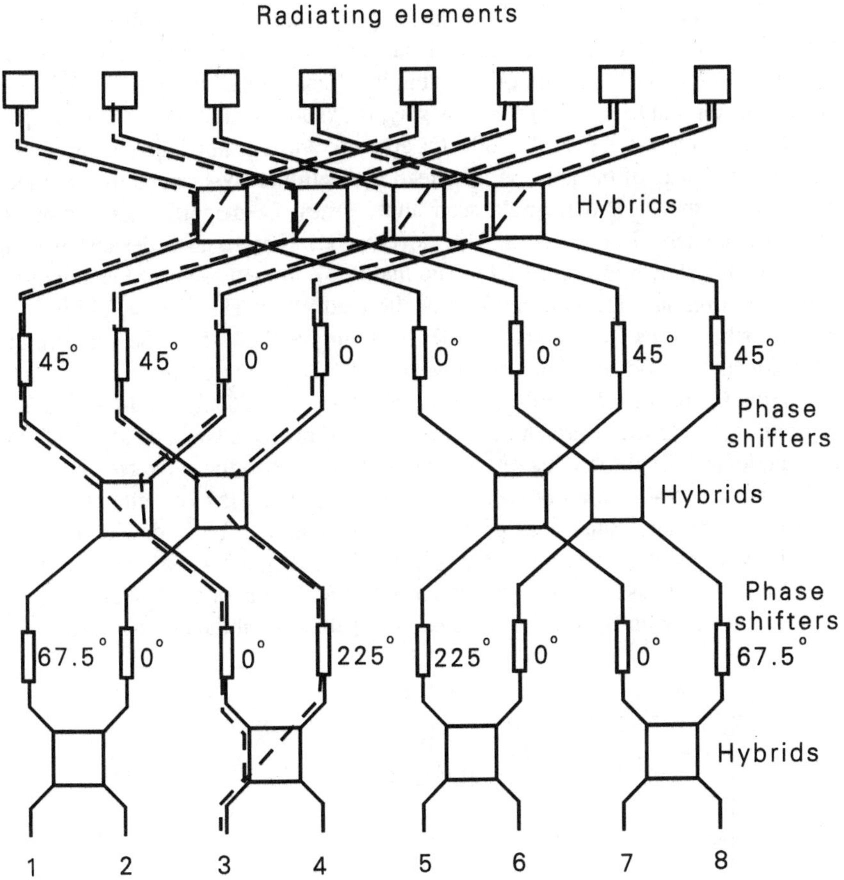

Radiating elements

Hybrids

45° 45° 0° 0° 0° 0° 45° 45°

Phase shifters

Hybrids

Phase shifters

67.5° 0° 0° 225° 225° 0° 0° 67.5°

Hybrids

1 2 3 4 5 6 7 8

Figure 3.16 Sketch of a Butler matrix network.

cylindrical arrays and has applications for a large shipboard UHF surveillance radar and electronically despun satellite antennas [52–54].

3.4 IMAGING REFLECTOR ANTENNAS

This section provides information on the single and the dual imaging reflector antenna for beam scanning [35].

3.4.1 Single-Reflector Imaging Antenna

A parabolic reflector antenna may be scanned through a limited off-axis angle by laterally displacing its feed about the boresight axis. However, as the feed moved off the boresight, phase aberration between the wavefronts arriving from different portions of the reflector increases. This effect degrades the beam, as has been reported by many authors [55, 56].

Figure 3.17 shows a sketch of the system in which the feed array is located between the main reflector and the focal plane. This way, the field distribution of the feed aperture is approximately recreated over the reflector aperture and a scan of the feed array results in a proportional scan of the antenna. The feed is electronically scanned by an angle that is equivalent to the desired far-field scan angle times the magnification factor of the imaging arrangement. LoForti et al. [57] developed a procedure to design an imaging reflector antenna shown in Figure 3.18. The main task in the design of the antenna is to determine the optimum location and size of the feed array. The feed should be kept as close as possible to the focal point in order to minimize the size of the aperture to that required to achieve good performance. However, a feed closer to the focus implies a larger magnification factor ($M = D/d$) of the system. Because the feed array must be scanned by an angle equivalent to the desired antenna scan angle times M, a larger M obviously means higher scan losses. Therefore, this sets a limit on how close to focus the feed array can be located and how small the array size can be. LoForti et al. [57] developed a software package using physical optics approximation to calculate the far field. Figures 3.19 and 3.20 show the computed radiation patterns in the vertical plane for an antenna with and without Taylor taper distribution. The layout and dimensions of the L-band antenna are shown in Figure 3.17 and Table 3.1. A tapered distribution on the circular array aperture

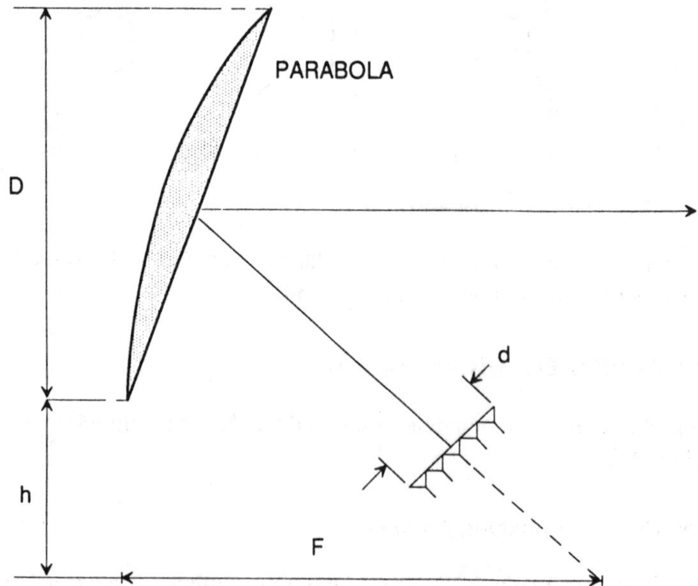

Figure 3.17 Sketch of a single-reflector imaging system [35].

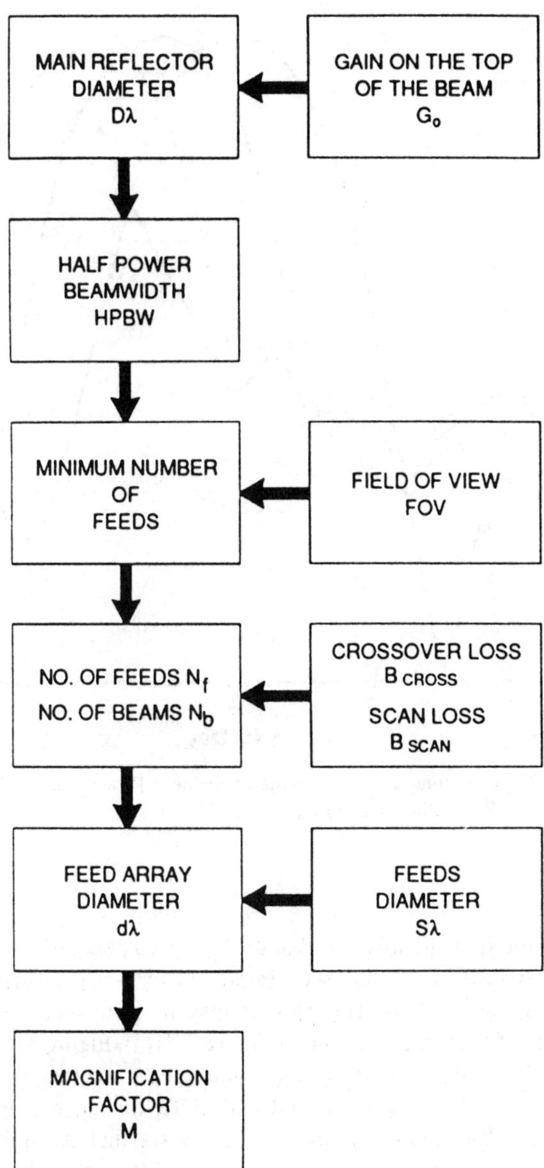

Figure 3.18 Design procedure of an imaging single-reflector antenna [35].

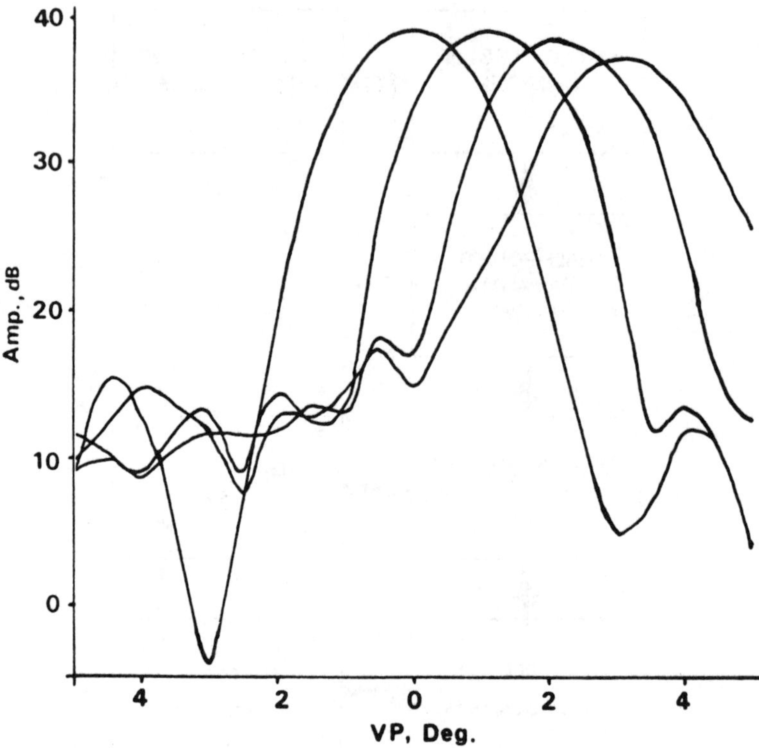

Figure 3.19 Single-reflector imaging antenna patterns of Figure 3.17 and Table 3.1 without Taylor taper distribution (VP = vertical plane) [35].

lowers the levels of the first sidelobes so that the specified copolarization isolation can be achieved. The first sidelobe level has been improved by 5 dB by using Taylor's distribution, as shown in Figure 3.20. The plot of loss in gain versus scan angle for the magnification factors of 5 and 9 are shown in Figure 3.21(a). Figure 3.21(b) shows the plot of HPBW/$HPBW_0$ versus the scan angle of the antenna for the magnification factors of 5 and 9. These curves are plotted without and with a Taylor taper distribution in Figures 2.21(a) and 2.21(b), respectively. Figure 2.21(a) shows that the gain is higher at the boresight of the antenna for the magnification factor of 9, but the loss in gain increases with the scan angle faster than that for the magnification factor of 5. In Figure 3.21(b) HPBW and $HPBW_0$ denote the half-power (3-dB) beamwidth of the array pattern after the magnification and half-power beamwidth of the array pattern before magnification *M*, respectively.

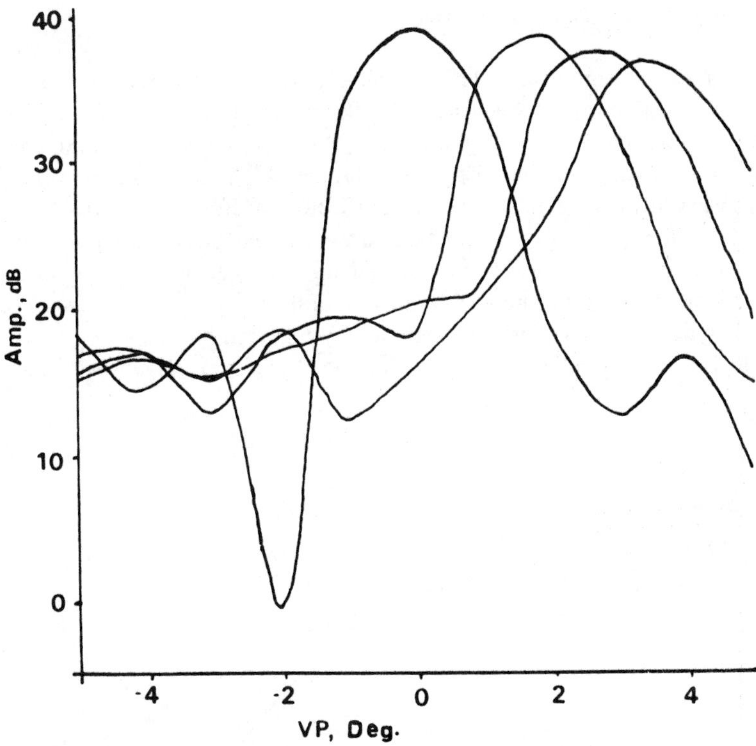

Figure 3.20 Single-reflector imaging antenna patterns of Figure 3.17 and Table 3.1 with Taylor taper distribution (VP = vertical plane) [35].

Table 3.1
Dimension of the Imaging Single Reflector Systems
(Figure 3.17)

$D = 8.82$ m
$h = 2.40$ m
$d = 1.764$ m
$M = 5$
Offset angle (Θ) = 43.57 deg
Feed horn aperture (single element) = 0.252m
Number of horns = 37

3.4.2 Dual-Reflector Imaging System

This antenna system comprises a large parabolic main reflector fed from a small linear phased array via a confocal Gregorian subreflector, as shown in Figure 3.22. The feed array produces an approximately plane wave, which is transformed into a converging spherical wave passing through F by the subreflector. After the focal point, the wave becomes spherically diverging and the main reflector transforms it to a plane wave. Beam scanning is performed by giving the feed array a linear phase variation. The imaging system inverts the scan direction so that an upward scan of the feed array gives a downward scan of the beam, and a leftward scan of the array gives a rightward scan of the beam. This type of antenna has been reported by many authors [58–74].

In this system, the phased array is magnified by the two reflectors and an image of

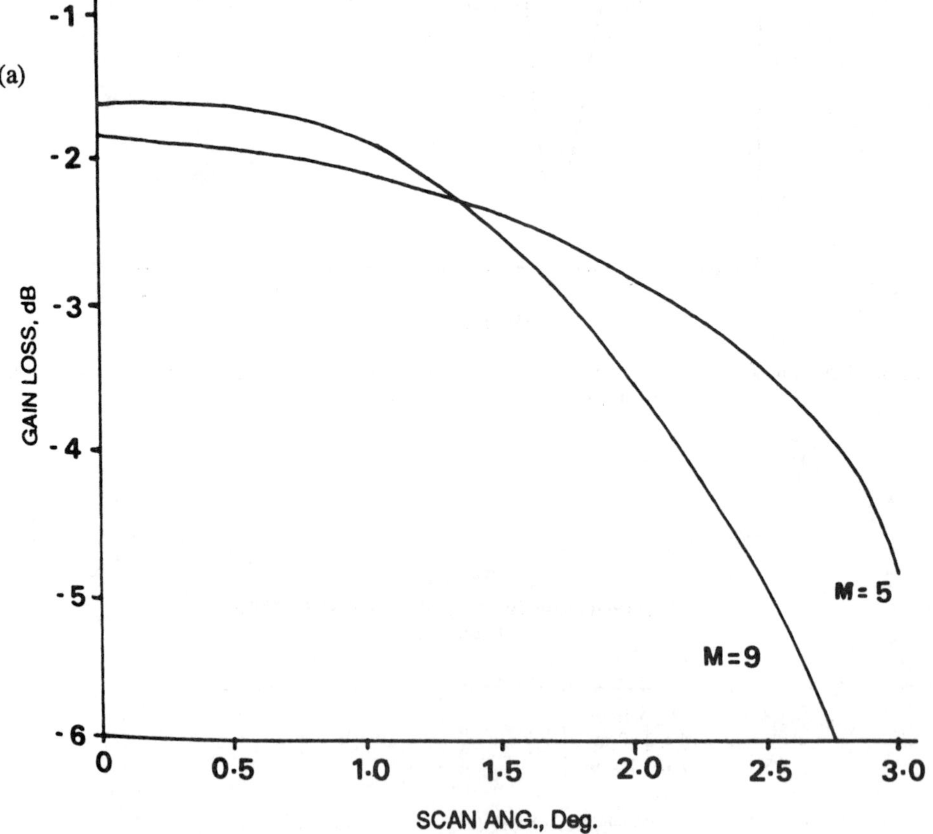

Figure 3.21 (a) Plot of loss in gain versus scan angle and (b) plot of half-power beamwidths versus scan angle for a single-reflector imaging antenna (Figure 3.17, Table 3.1) of magnification factors of 5 and 9 (no taper distribution) [35].

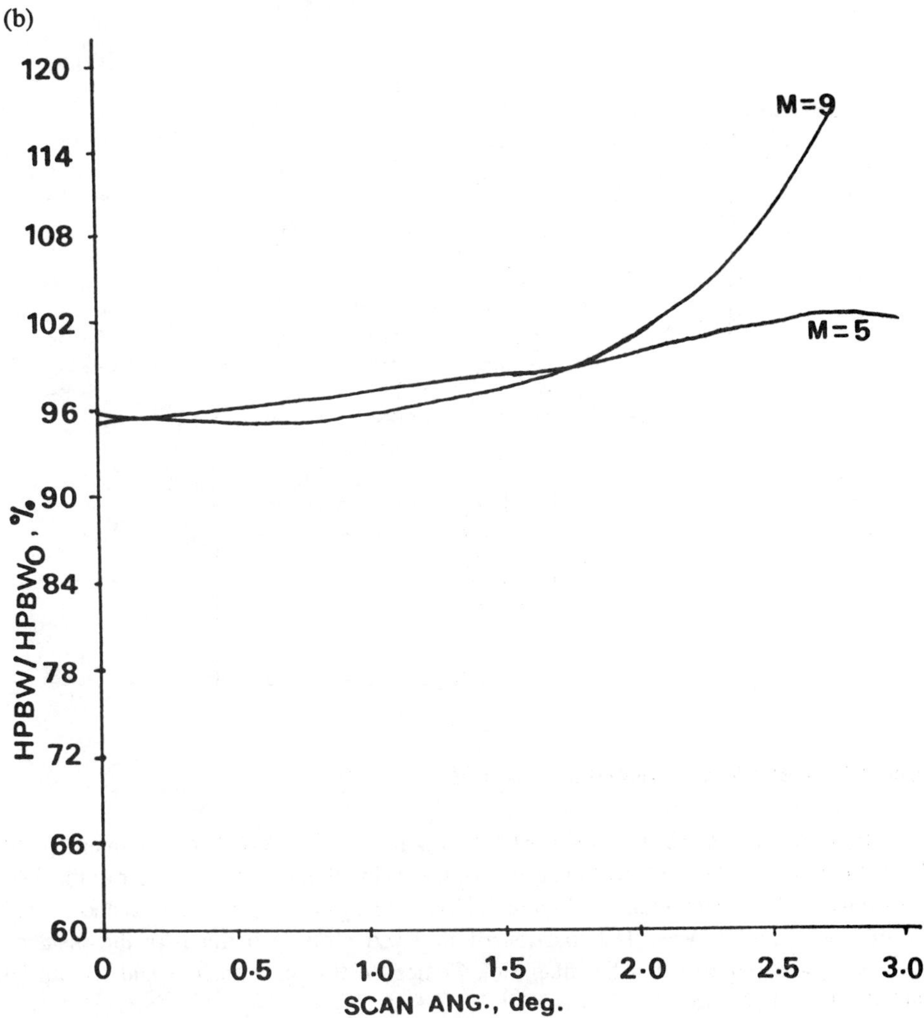

Figure 3.21 (continued)

a near field of the phased array is formed on the aperture of the main reflector. The magnification factor, *M,* for the antenna is defined as the ratio of the sizes of the main reflector aperture and the array aperture and is denoted as

$$M = \frac{D_2}{D_1} = \frac{f_2}{f_1} \tag{3.9}$$

where D_1 denotes the diameter of the feed array, D_2 the diameter of the main reflector, f_2 the axial focal length of the main reflector, and f_1 the axial focal length of the subreflector.

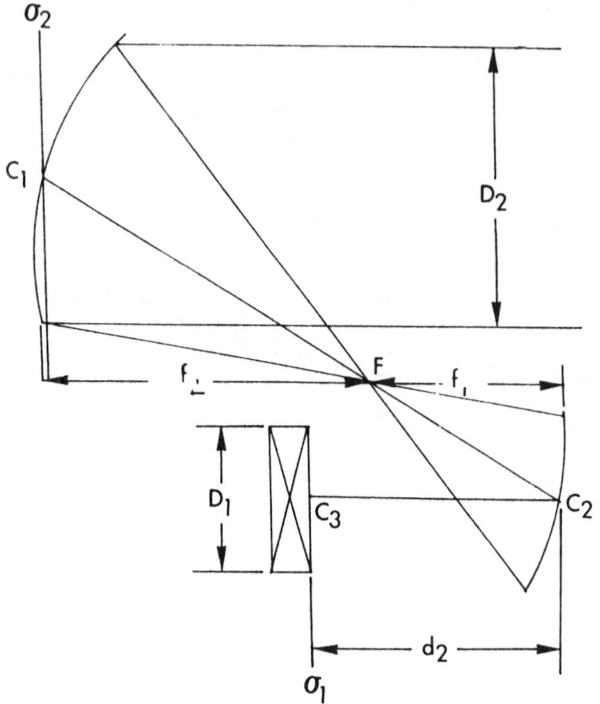

Figure 3.22 Dual-reflector imaging antenna system [35].

By choosing $f_2/f_1 >> 1$, a small feed array diameter (D_1) is sufficient to intercept all the incident rays. We assume that the center of the main reflector and the center of the feed array must be conjugate points. In Figure 3.22 C_1 and C_2 are conjugate points and σ_1 and σ_2 are conjugate surfaces. The location of the feed array is defined by the distance $C_2C_3 = d_2$ and requires that the image of C_3 lies on the main surface and d_2 can be calculated from [35] as

$$d_2 = FC_1 \frac{M+1}{M} \tag{3.10}$$

where F is the common focal point as shown in Figure 3.22. From Brueggemann [75], we can say that for each of the feed element positions, there is a corresponding conjugate point on the output ray. These conjugate points lie on a surface given by [74] as

$$Z_c = 2(M+1)\frac{x^2 + y^2 - (M \cdot d)^2}{4f_0} + \frac{(M \cdot d)^2}{4f_0} \tag{3.11}$$

in an x,y,z coordinate system with origin in the paraboloidal main reflector vertex and d is the distance from the axis to the center feed element, as shown in Figure 3.23. Expression (3.11) shows that the surface of conjugate points is a paraboloid with a focal length of $f_0/[2(M + 1)]$.

Computed Results

Figure 3.24 shows a dual-reflector imaging system design by LoForti et al. [57] at Telespazio with a 44.1λ main reflector. The magnification factor is 5, and the feed array is a 37-element hexagonal array [74]. The element feed is a model of a circular waveguide horn with a TE_{11} mode. The feed radius is $r_f = 0.63\lambda$. Two sets of feed excitations are considered—one with equal amplitude distribution and one with a tapered distribution.

Figures 3.25 and 3.26 show the computed radiation pattern by the dual *physical optics* (PO) method for the dual-reflector system shown in Figure 3.24. The agreement is excellent between the results of Albertsen et al. [74] and LoForti et al. [57]. Both results are computed with the PO method.

Crescimbeni et al. [59] reported some computed results for a dual-imaging reflector antenna of a magnification factor of 3, and the remaining parameters are the same as in Figure 3.24. Figure 3.27 shows gain loss and BW/BW_0 versus the scan angle for different

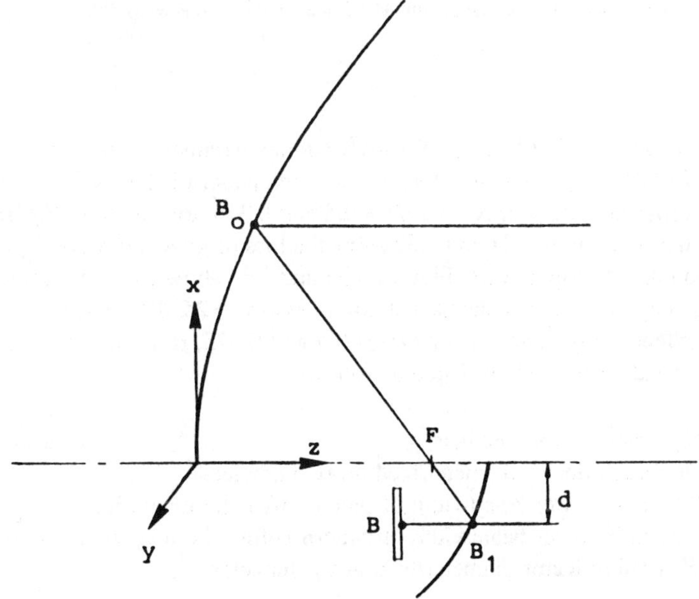

Figure 3.23 Coordinate system of an imaging dual-reflector antenna [35].

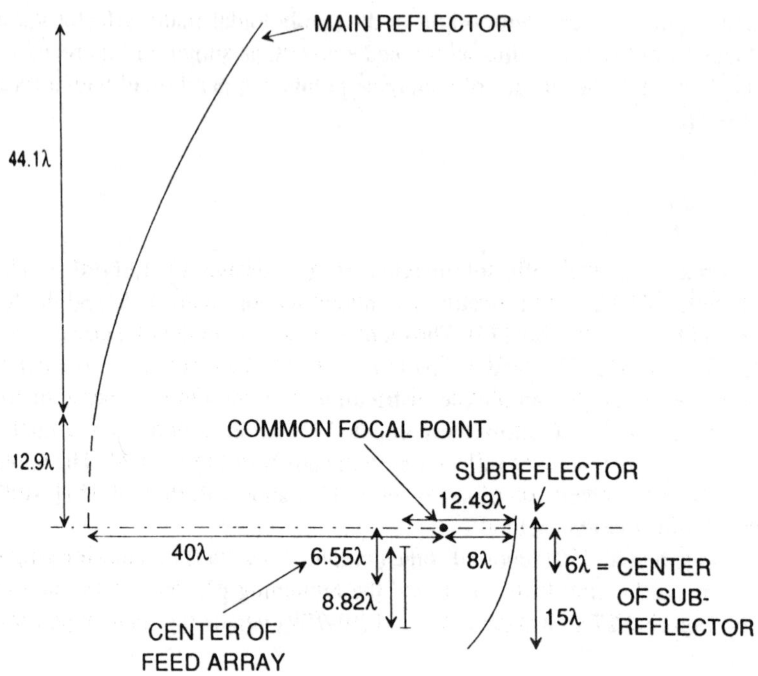

Figure 3.24 Imaging dual-reflector design with 44.1 λ main reflector diameter [35].

R values and F/D = 0.5. Figure 3.28 shows similar parameters for F/D = 0.375. First sidelobe and $\Theta/BW_{-3\text{ dB}}$, beyond which the antenna pattern is below −25 dB versus the scan angle; different values of R for F/D = 0.5 and 0.375 are shown in Figures 3.29 and 3.30, respectively. Figures 3.31 and 3.32 show the loss of gain and BW/BW_0 versus r_1 for F/D = 0.5 and 0.375, respectively. Figures 3.33 and 3.34 show the first sidelobe level and $\Theta/BW_{-3\text{ dB}}$, beyond which antenna pattern is below −25 dB versus the subreflector diameter (r_1)/feed array diameter for F/D = 0.5 and 0.375, respectively. Various parameters in Figures 3.27 to 3.34 are listed as follows:

$G_{\text{imaging}} - G_0$ = gain loss in decibels
R = subreflector diameter/feed array diameter
BW = half-power beamwidth of pattern after the magnification
BW_0 = half-power beamwidth of pattern before the magnification M
R_1 = subreflector diameter/feed array diameter

In Figures 3.31, 3.32, 3.33, and 3.34, r denotes different main reflector oversizing values.

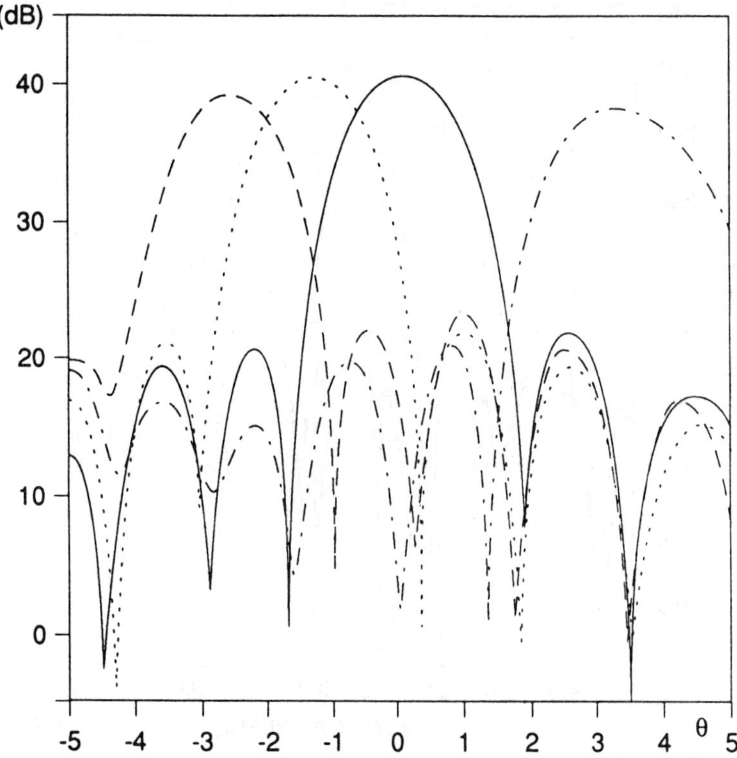

Figure 3.25 Computed patterns for the scan angles 0°, −1.5°, −3°, and 3° plotted by Albertsen et al. [74] without array taper.

3.5 RF ELECTRONIC BEAM-STEERING TECHNIQUES FOR OPTICAL FIBERS

This section provides a short review of RF/microwave beamforming/steering techniques that will be used for optical fibers.

One of the common techniques of electronic beam steering in array antennas is that of phase scanning in which the RF/microwave signal from a source is split by means of power dividers and fed through different lengths of lines and switches to create phase shift to individual radiating elements as shown in Figure 3.35. This technique for the variation of phase is bulky and has very limited beam steering at microwave frequencies. However, this idea has been used to make delay lines at optical frequencies, which will be discussed in Chapters 5, 6, and 7. Another technique for beam steering is achieved by means of control signals applied to the phase shifters, which then causes an approximate linear phase variation of the RF/microwave signal across the array as shown in Figure 3.36. In

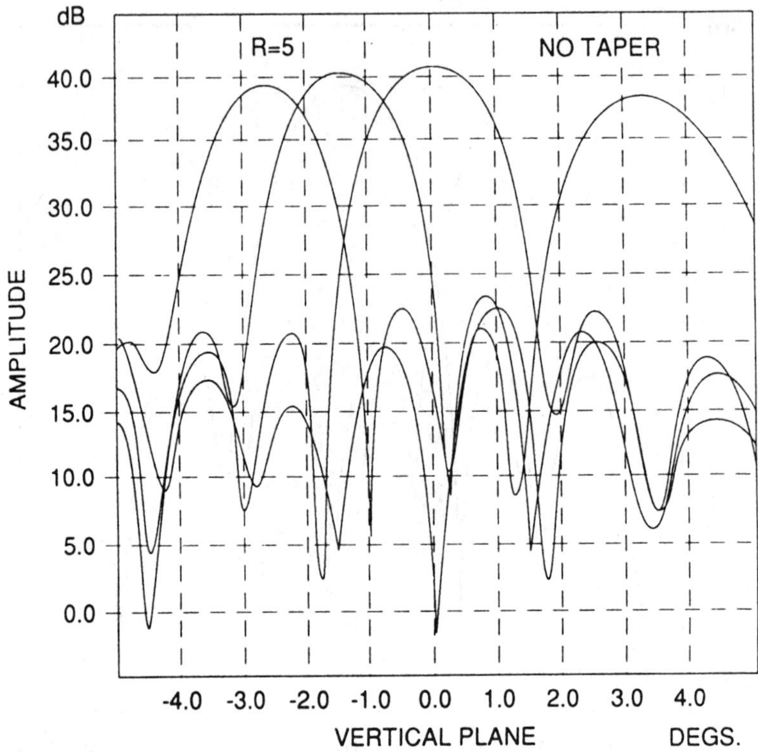

Figure 3.26 Computed patterns for the scan angles 0°, −1.5°, −3°, and 3° plotted by LoForti et al. [47].

both techniques, the linear phase variation provides a wavefront in the required angular direction to steer the beam.

Figure 3.36 shows a receiving-type linear array antenna. The normalized electromagnetic field vector summation of signals from antenna elements is given by

$$E = \sum_{i=0}^{i=n} \sin\left\{\omega\left[1 + \frac{D_i}{C_1} + \frac{l_i}{C_2}\right]\right\} \tag{3.12}$$

where $\omega = 2\pi f$ denotes angular frequency in radians, C_1 is the velocity of propagation of electromagnetic waves D_i, C_2 is the velocity of propagation of the electromagnetic wave l_i, d_i is the distance from the 0th to ith antenna elements, l_i is the distance from ith antenna element to the wavefront as shown in the figure and $l_i = d_i \sin\Theta$. Θ denotes the wavefront angle-of-arrival of the electromagnetic wave, D_i the distance of RF/microwave line to/from the source to the input of antenna element, $(T_1)_i = D_i/C_1$, and $(T_2)_i = l_i/C_2$.

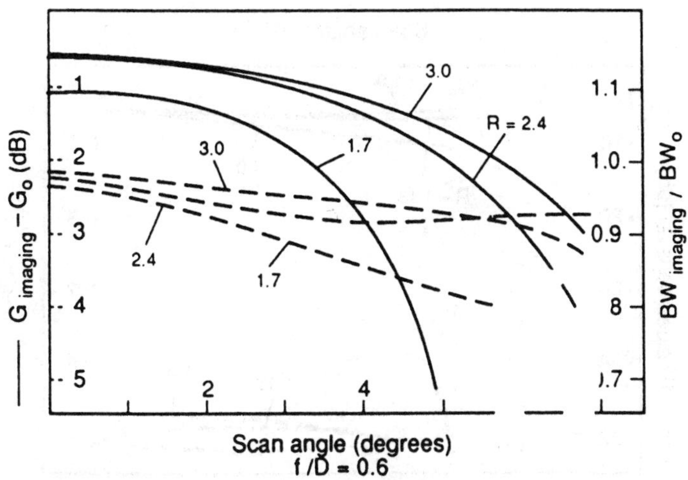

Figure 3.27 Loss of gain and BW/BW_0 versus scan angle for different R values and $F/D = 0.5$ [35].

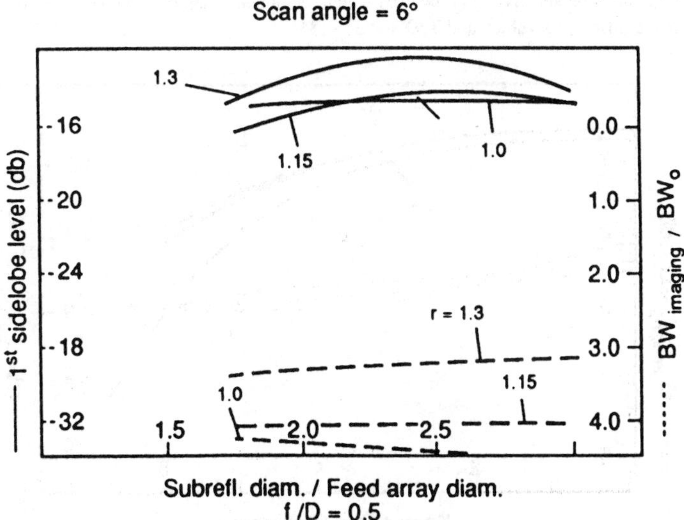

Figure 3.28 First sidelobe and $BW_{-3\ \mathrm{dB}}$, beyond which the antenna radiation pattern is below −25 dB versus R for different r values and $F/D = 0.5$ [35].

Equation (3.12) can be rewritten as

$$E = \sum_{i=0}^{i=n} \sin \{\omega [t + (T_1)_i + (T_2)_i \sin \Theta]\} \tag{3.13}$$

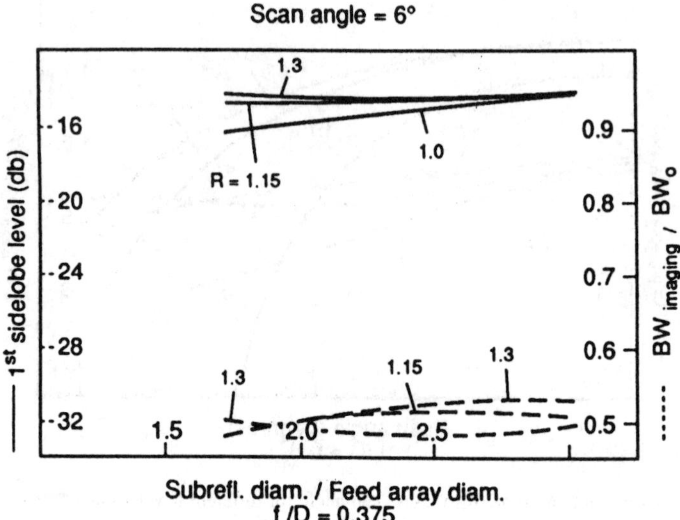

Figure 3.29 First sidelobe and $BW_{-3\ \text{dB}}$, beyond which the antenna radiation pattern is below −25 dB versus R for different r values and F/D = 0.375 [35].

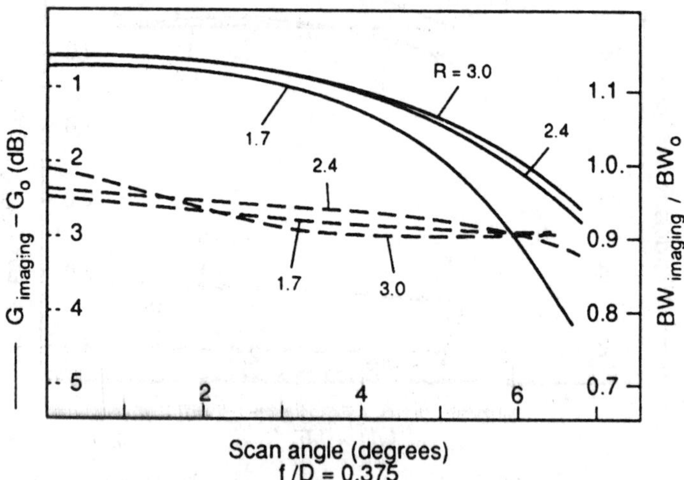

Figure 3.30 Loss of gain and BW/BW_0 versus scan angle for different R values [35].

When the path delays to the summing point via each antenna element are equal, the sine functions in the summation of (3.13) will be added in phase for all values of frequency f, thereby forming a summation beam in the direction Θ, which is independent of frequency. This condition is given by

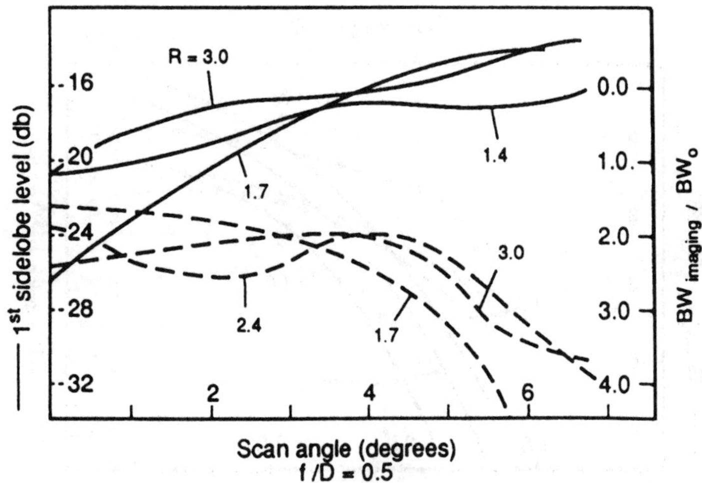

Figure 3.31 First sidelobe level and $BW_{-3\ dB}$, beyond which the antenna pattern is below −25 dB versus scan angle for different R values [35].

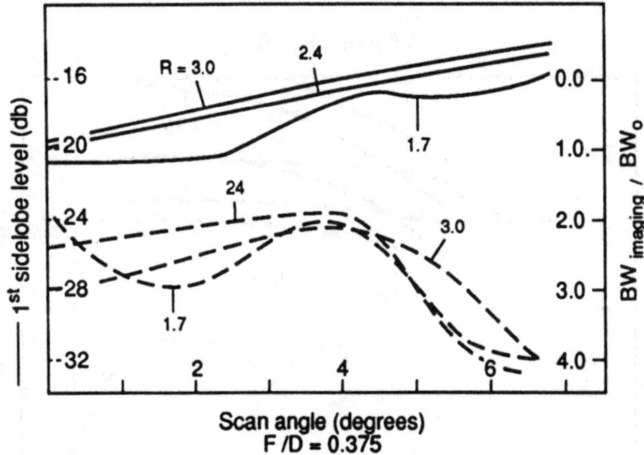

Figure 3.32 First sidelobe level and $BW_{-3\ dB}$, beyond which the antenna pattern is below −25 dB versus scan angle for different R values [35].

$$(T_1)_i + (T_2)_i \sin \Theta = \text{constant} \tag{3.14}$$

The antenna beam formed in the Θ direction is steered by controlling the relative lengths of D_i and adding parallel summing networks from each antenna element.

For a large array, the overall system becomes very complex; therefore optical fiber delay compressive delay line is the best solution to compress the overall hardware com-

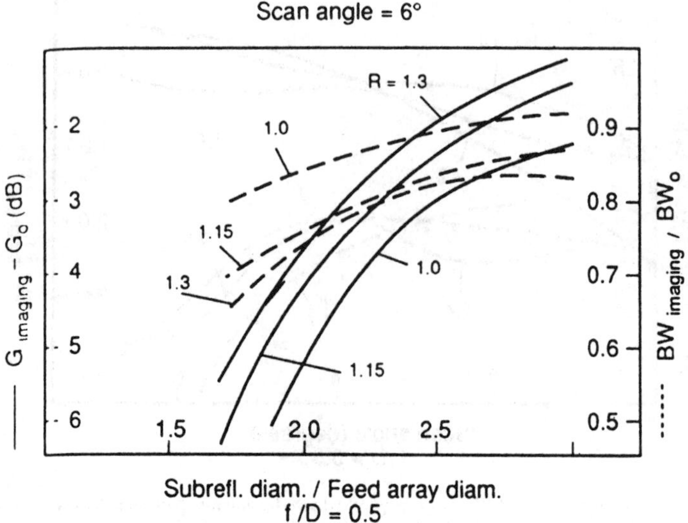

Figure 3.33 Loss of gain and BW/BW_0 versus R for different main reflector oversizing (r) values [35].

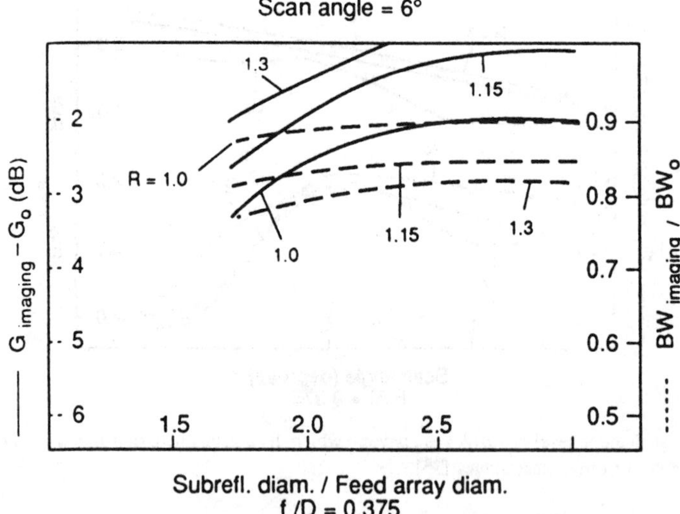

Figure 3.34 Loss of gain and BW/BW_0 versus R for different main reflector oversizing (r) values [35].

plexity with respect to both the number of delays per element and the number of elements per phased array antenna. A description of hardware-compressive fiber optic delay line architecture for the time steering of a phased array antenna is described in Chapter 6.

Figure 3.37 shows the current technique for an electronic beamforming network at RF/microwave frequencies [76]. In the figure, each T/R unit is connected to a 6-bit

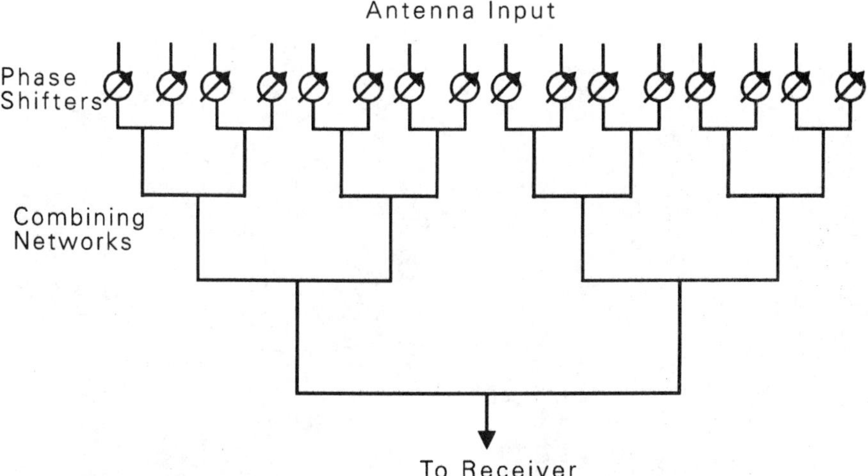

Figure 3.35 RF electronic beam steering using phase shifters.

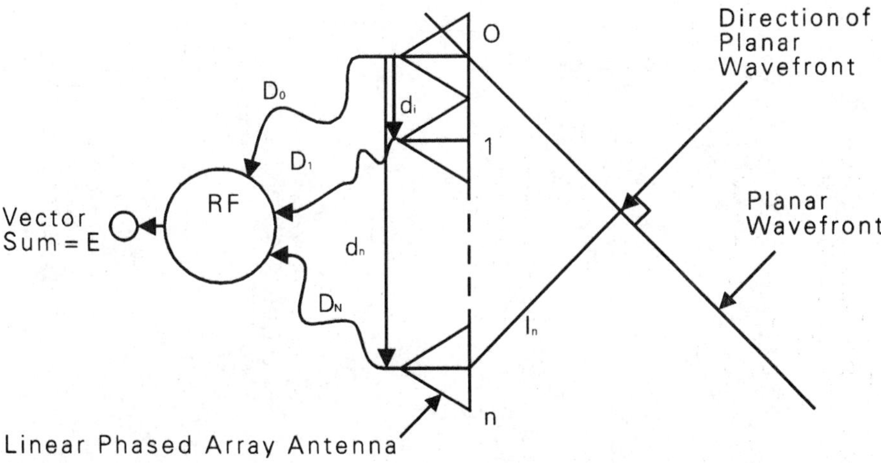

Figure 3.36 Concept of true-time delay beamforming.

microwave time delay unit and a 1 × 8 RF/microwave power combiner/divider. The output of the power combiner is the received signal and input to the divider is the transmit signal.

In 1928, small antennas with electron tube amplifiers were used in radio broadcast receivers at 1 MHz. This type of device was named as an active antenna. Adler [77] developed an expression for the frequency range over which an oscillator remains locked in phase to an injected signal. In 1962, Mackey [78] extended Adler's analysis to include effects of phase modulation on the injected signal and reported that injection-locked

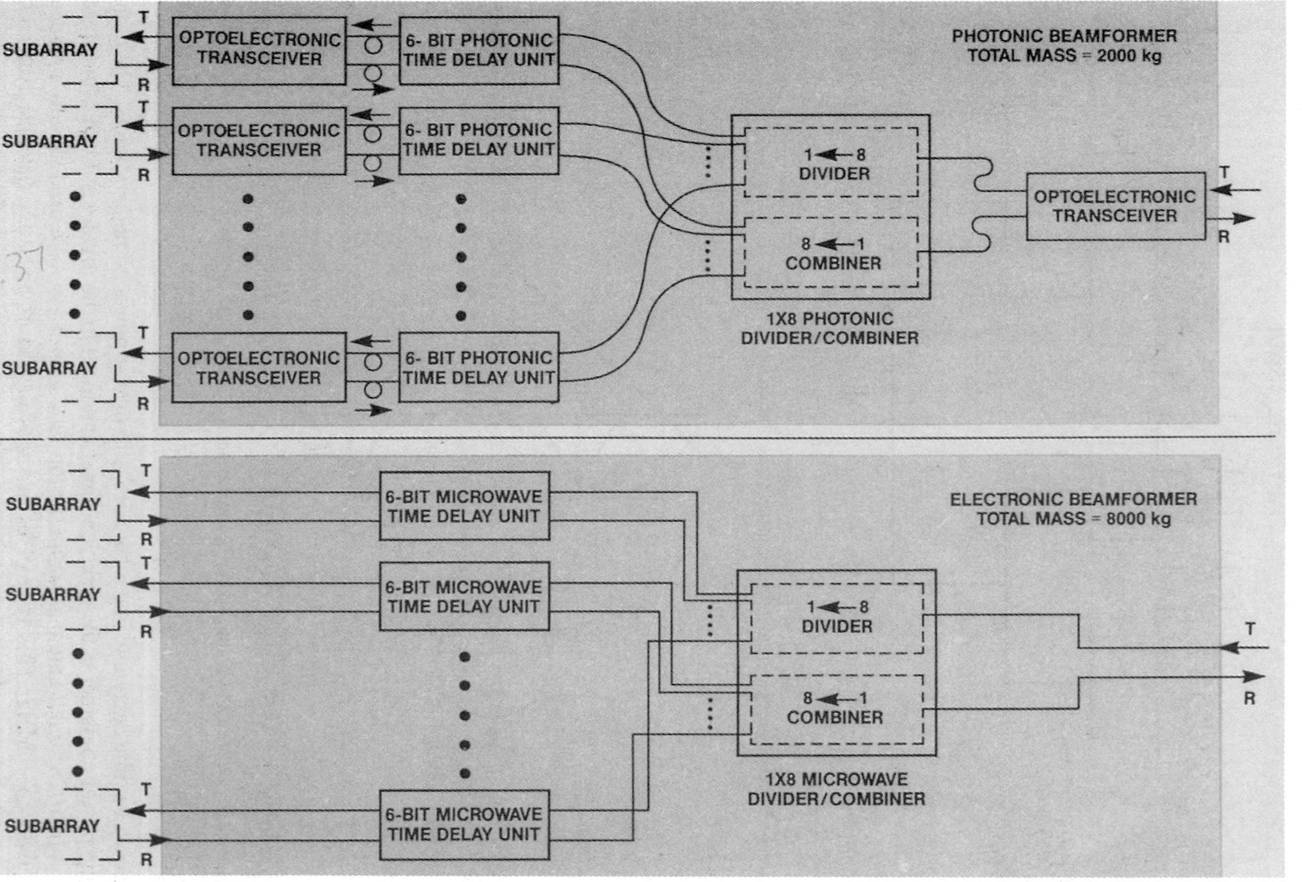

Figure 3.37 The current architecture for an electronic true-time delay beamformer [76].

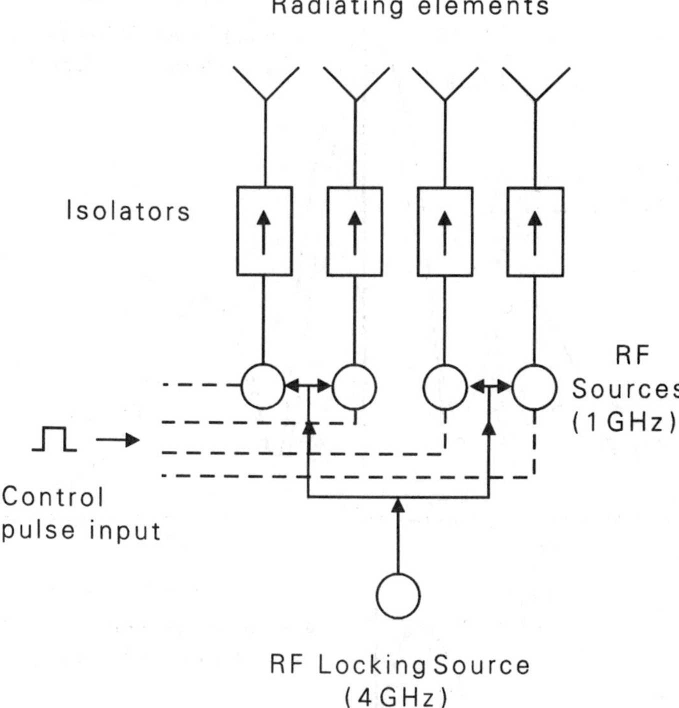

Figure 3.38 Illustration of a four-element array with injection locking. (© 1974 IEEE.)

amplifiers had some advantage over conventional amplifiers in the RF/microwave frequency band. Kurokawa [79] reported injection locking of RF/microwave solid-state oscillators. Since the 1970s work with active array antennas has received much attention. Active antennas have been developed to integrate the oscillator circuit with passive antennas, thereby reducing transition and transmission losses [80]. Al-Ani et al. [81] reported a method of beam steering for an active array using locked oscillators. Passive phase shifters are not needed since, by locking the oscillators with a reference signal harmonically related to the fundamental output, a dual function of locking and phase shifting may be achieved. Each individual RF/microwave source is phase locked by a stable locking signal that is close in frequency to the nth harmonic of its free-running frequency. The fundamental output frequency of the RF/microwave source is shifted in phase by increments of $2\pi/n$, by applying pulses of appropriate amplitude and duration to the dc circuitry. Al-Ani et al. constructed and tested a four-element array (Figure 3.38) at 1 GHz using BFY 90 transistor (type of transistor) oscillators locked at 4 GHz to provide a 90° phase increment. Figure 3.39(a) shows the measured and calculated radiations of the broadside position with all elements in phase. Figure 3.39(b) shows the steered beam

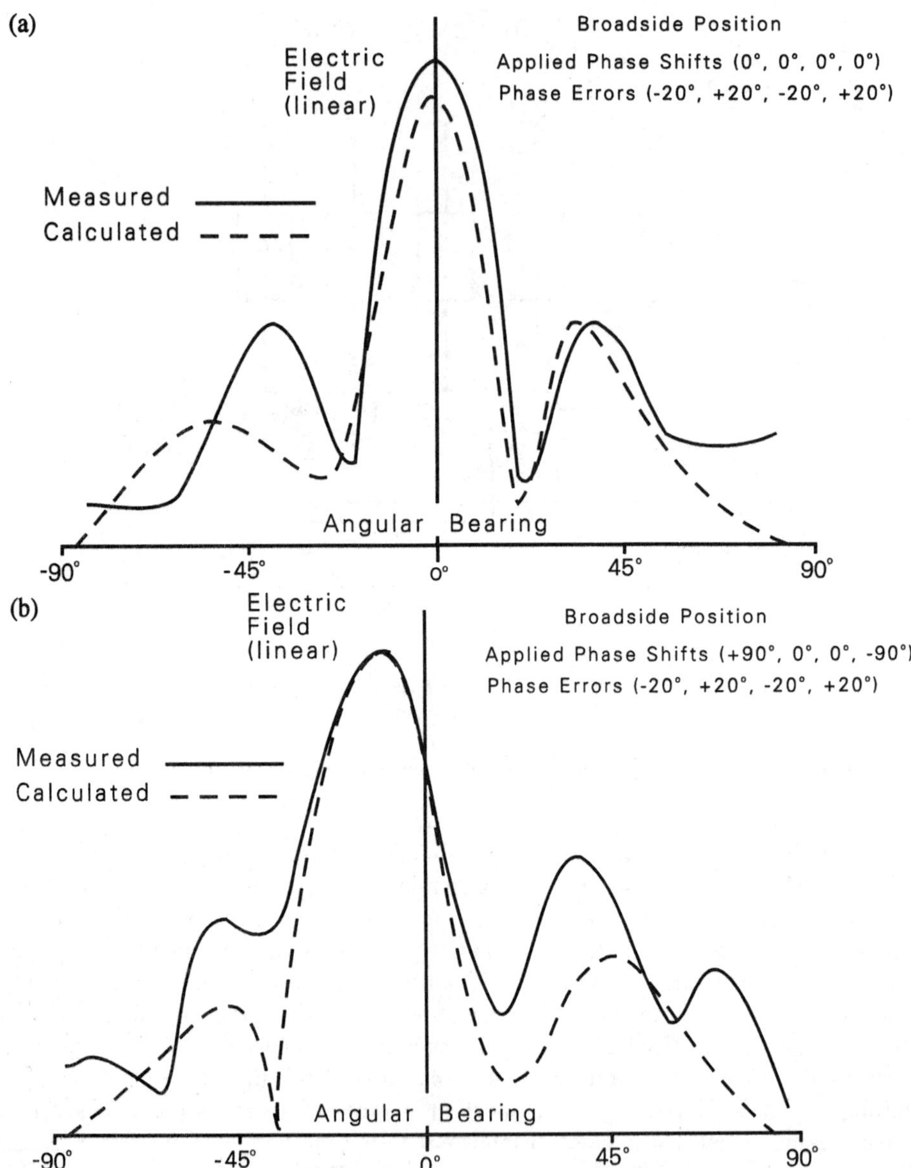

Figure 3.39 Antenna radiation pattern: (a) broadside position and (b) shifted position. (© 1974 IEEE.)

pattern with the end elements radiating with phase shift of ±90°. The main disadvantage of this technique is the need to supply locking power at a harmonic of the fundamental output frequency instead of at the fundamental frequency itself, together with associated microwave circuitry at the harmonic frequency. The amount of harmonic locking power that is needed is a strong function of the nonlinear active device characteristics, and it is found in the case of transistor oscillators that the locking power required at the fourth harmonic is about 10 dB greater than that required at fundamental frequency. The locking power will be higher for higher harmonics, and it is therefore felt that the method has applications for low-order harmonics ($n < 4$) and for radar systems with fundamental frequencies at or below X-band. This limits the phase shift increment to 90° in case of the fourth harmonic ($2\pi/n$). Stephan [82] used the interinjection-locked oscillators for power combining and steering of an active phased array antenna. He proposed a linear phased array driven by interinjection-locked oscillators as shown in Figure 3.40. The maximum one-sided beam-steering angle is given by

$$\Gamma_m = \sin^{-1}\left(\frac{\Theta_m}{\phi}\right) \tag{3.15}$$

where Θ_m is the maximum phase progression per element and ϕ is the element separation in electrical degrees. As an example, the element spacing of one-half free-space wavelength ($\phi = 180°$) and $\Theta_m = 60°$ gives a beam steering of ±20°. The interinjection-locked concept can be used for the linear receiving phased array as well as for two-dimensional arrays.

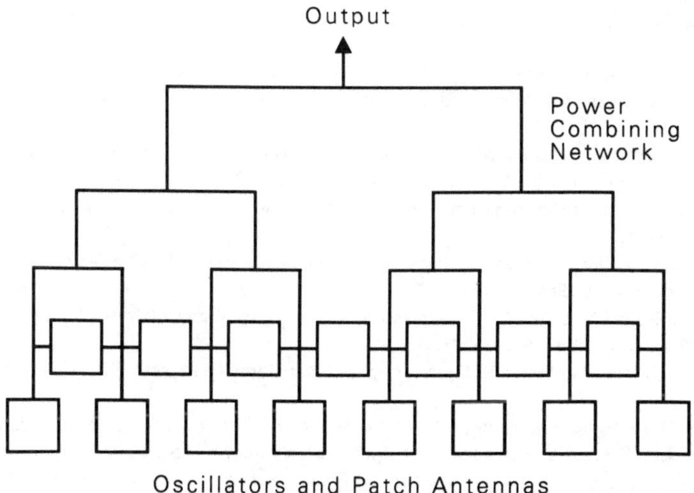

Figure 3.40 A linear phased array driven by interinjection locked oscillators. (© 1986 IEEE.)

Recently, Navarro et al. [83] reported electronic beam steering of active antenna arrays. They use DC bias to change the self-oscillating frequencies of individual active antenna elements and imposed an injection-locking phase shift between active antenna elements. A 2×2 active antenna array provided 15° of electronic beam steering and less than a 1.2-dB loss in output power. The technique can be used for several antennas if a ramped DC voltage is applied.

References

[1] Luneburg, R. K., *Mathematical Theory of Optics,* University of California Press, 1964.

[2] Gutman, A. S., ''Modified Luneburg Lens,'' *J. Appl. Phys.,* Vol. 23, No. 7, 1954, p. 855.

[3] Morgan, S. P., ''General Solution of the Luneburg Lens Problem,'' *J. Appl. Phys.,* Vol. 29, No. 9, 1958, p. 1358.

[4] Cassel, K. E., ''Broad Band ECM/ESM Antenna with 360 Degrees Multiple Beam Coverage,'' *Military Microwaves Conf.,* London, U.K., October 1982, pp. 512–518.

[5] Vogel, M., ''New Kind of Planar Waveguide Luneburg Lens for the mm Wave Region,'' *13th European Microwave Conf.,* Numberg, 1986, pp. 413–418.

[6] DuFort, E. C., ''Wide Angle Scanning Optical Antenna,'' *IEEE Trans. Ant. Prop.,* Vol. AP-131, No. 1, 1983, pp. 60–67.

[7] Wild, J. P., Geodesic Lens, Australian Patent, 1976, Reference No. AU-B 20708/76.

[8] Ruze, J., ''Wide-Angle Metal-Plate Optics,'' *Proc. IRE,* Vol. 38, 1950, pp. 53–69.

[9] Rotman, W., and R. F. Turner, ''Wide Angle Microwave Lens for Line Source Applications,'' *IEEE Trans. Ant. Prop.,* Vol. AP-11, 1963, pp. 623–632.

[10] Niazi, A. Y., ''Rotman Lens Fed Multiple Beam Array,'' *2nd IEE Int. Conf. Ant. Prop.,* April 1981, York, U.K., pp. 93–97.

[11] Smith, M. S., ''Rotman Lens Multiple Beamformers,'' *Military Microwave Conf.,* Brighton, U.K., 1986, pp. 279–289.

[12] Smith, M. S., ''Design Consideration of Ruze and Rotman Lens,'' *Radio & Elect. Eng.,* Vol. 52, No. 4, 1982, pp. 181–187.

[13] Jasik, H. (ed.), *Antenna Engineering Handbook,* New York: McGraw-Hill, 1961.

[14] Archer, D. H., ''Lens-Fed Multiple Beam Array,'' *Microwave J.,* Vol. 27, No. 9, September 1984, pp. 171–195.

[15] Kumar, A., *Metallic and Dielectric Lens Antennas and Their Applications,* Research Report, AK-EETC-UK-02–88, AK Electromagnetique Inc., Quebec, Canada, 1988.

[16] Smith, M. S., ''Design Considerations for Ruze and Rotman Lens,'' *Radio & Elect. Eng.,* Vol. 52, No. 4, 1982, pp. 181–187.

[17] Butler, J., and R. Howe, ''Beamforming Matrix Simplifies Design of Electronically Scanned Antennas,'' *Elec. Design,* Vol. 9, 1961, pp. 170–173.

[18] Fong, A. K. S., and M. S. Smith, ''Design and Performance of a Microstrip Multiple Beam Forming Lens,'' *Third IEE Int. Conf. Ant. Prop.,* 1983, pp. 344–347.

[19] Kumar, A., *Design of a Microstrip Rotman Lens Antenna,* Research Report EETC, Coventry, U.K., January 1982.

[20] Mailloux, R. J., ''Hybrid Antennas,'' *Handbook of Antenna Design* (A. W. Rudge, K. Milne, A. D. Olver, and P. Knight, Eds.), IEE Electromagnetic Wave Series, London, U.K.: Peter Periginus, 1982.

[21] Smith, M. S., and A. K. S. Fong, *Amplitude Performance Rotman Lens,* Research Report, 1982.

[22] Maybell, M., ''Printed Rotman Lens-Fed Array Having a Wide Bandwidth, Low Sidelobes, Constant Beamwidth and Synthesised Radiation Pattern,'' *IEEE AP-S Symp.,* Houston, Texas, 1983, pp. 373–376.

[23] Herd, J. S., and D. M. Pozar, "Design of a Microstrip Array Fed by a Rotman Lens," *IEEE AP-S Symp.*, Boston, MA, 1984, pp. 729–732.

[24] Kumar, A. "Microstrip Rothman Lens Beamforming for Satellite Antennas," *IEEE Montech'86 Conf. Ant. and Comm.*, Montreal, Quebec, 1986.

[25] Musa, L., and M. S. Smith, "Microstrip Lens Port Design," *IEEE AP-S Symp.*, Philadelphia, PA, June 1986.

[26] Parini C. G., and L. M. C. E. Lee-Yow, "Performance of Waveguide Rotman Lens Beamforming Networks in the Presence of Mutual Coupling," *Int. Conf. Ant. Prop.*, York, 1987, pp. 1153–1156.

[27] Lee, J. J., "Lens Antennas," *Antenna Handbook Theory Applications and Design* (Y. T. Lo, and S. W. Lee, Eds.), New York: Van Nostrand Reinhold, 1988, Ch. 16.

[28] Cornbleet, S., *Microwave Optics*, London, U.K.: Academic Press, 1976.

[29] Poulton G. T., "Wide-Angle Lens Antennas for Interscan," *IREECON Convention Digest*, 1977, pp. 24–27.

[30] Rogers, P. G., "Design of Compact Low Loss Rotman Lenses," *IEE Proc.*, Pt. H, Vol. 134, No. 5, 1987, pp. 449–459.

[31] Sole G. C., and M. S. Smith, "Multiple Beam Forming for Planar Antenna Arrays Using a Three-Dimensional Rotman Lens," *IEE Proc.*, Pt. H, Vol. 134, No. 4, August 1987, pp. 375–385.

[32] Luh, H. S., T. M. Smith, and W. G. Scott, "Dual Band TEM Lens Development," *IEEE AP-S Symp.*, May 1978, pp. 369–373.

[33] Rappaport, C. M., "Optimised Three-Dimensional Lenses for Wide-Angle Scanning," *IEEE Trans. Ant. Prop.*, Vol. AP-23, November 11, 1985, pp. 1227–1236

[34] McGrath, D. T., "Planar Three Dimensional Constrained Lenses," *IEEE Trans. Ant. Prop.*, Vol. AP-34, No. 1, January 1986, pp. 46–49.

[35] Kumar, A., *Fixed and Mobile Terminal Antennas*, Norwood, MA: Artech House, 1991.

[36] DeVore, H. B., and H. Iams, "Microwave Optics Between Parallel Conducting Sheets," *RCA Rev.*, Vol. 9, 1948, pp. 721–732.

[37] Boyns, J. E., A. D. Munger, J. H. Provencher, J. Reindel, and B. I. Small, "A Lens Feed for a Ring Array," *IEEE Trans. Ant. Prop.*, Vol. AP-16, March 1968, pp. 264–268.

[38] Bodnar, D. G., B. K. Rainer, and Y. Rahmat-Samii, "Lens Antenna Concept for Land Mobile Satellite Communications," *Proc. IEEE Vehicular Tech. Conf.*, Dallas, IEEE Cat. No. 86CH2308-5, May 1986, pp. 8–14.

[39] Bodnar, D. G., B. K. Rainer, and J. A. Cribbs, "Lens Feed Array for Mobile Satellite Applications," Final Technical Report GTRI Project A-4263, JPL Contract 957307, Georgia Inst. of Tech., Atlanta, GA, June 1986.

[40] Bodnar, D. G., B. K. Rainer, and Y. Rahmat-Samii, "A Novel Array Antenna for MSAT Applications," *IEEE Trans. Vehicular Technology*, Vol. 38, No. 2, May 1989, pp. 86–93.

[41] Blass, J., "Multi-Directional Antennas—New Approach Top Stacked Beams," *IRE Int. Convention Record*, Pt 1, 1960, pp. 48–50.

[42] Vetterlein, S. J., and P. S. Hall, "Novel Multiple Beam Microstrip Patch Array with Integrated Beamformer," *Electron. Lett.*, Vol. 25, No. 17, 1989.

[43] Shelton, J. P., and K. S. Kelleher, "Multiple Beams for Linear Arrays," *IRE Trans. Ant. and Prop.*, Vol. 9, March 1961, pp. 154–161.

[44] Muenzer, P. J., "Properties of Linear Phased Arrays Using Butler Matrices," NTZ, Heft 9, 1972, pp. 419–422.

[45] Macnamara, T., "Simplified Design Procedure for Butler Matrices Incorporating 90 DEG. or 180 DEG. Hybrids," *IEE Proc.*, Pt. H, Vol. 134, No. 1, 1987, pp. 50–54.

[46] Guy, J. R. F., "Proposal to Use Reflected Signals Through a Single Butler Matrix to Produce Multiple Beams from a Circular Array Antenna," *Electron. Lett.*, Vol. 28, No. 5, 1985, pp. 209–211.

[47] Blokhia, N. A., and B. A. Mishustin, "Design of Planar Beam-Shaping Circuits," *Radio Electron. Comm. Systems*, Vol. 27, No. 2, 1984, pp. 45–48.

[48] Chow, P. E. K., and D. E. N. Davis, "Wide Bandwidth Butler Matrix Network," *Electron. Lett.,* Vol. 20, No. 3, 1967, pp. 252–253.

[49] Withers, M. J., "Frequency Insensitive Phase Shift Networks and Their Application in a Wide Bandwidth Butler Matrix," *Electron. Lett.,* Vol. 5, No. 20, pp. 496–498.

[50] Levy, R., "A High Power X-Band Butler Matrix," *Microwave J.,* Vol. 27, No. 4, April 1984, pp. 135–140.

[51] Wallington, J. R., "Analysis Design and Performance of a Microstrip Butler Matrix," *1973 European Microwave Conf.,* Brussels, September 1973, pp. A14.3.1–A14.3.4.

[52] Small B. I., E. A. Killick, and J. Croney, "A Cylindrical Array for Electronic Scanning," *European Microwave Conf.,* London, September 1969, pp. 133–136.

[53] Chang, B., R. Yaminy, and R. Jackson, "A Foam Dielectric Matrix Fed Electronically Despun Circular Array," *IEEE AP-S Symp.,* September 1970, pp. 29–36.

[54] Sheleg, B., "Butler Submatrix Feed Systems for Antenna Array," *IEEE Trans. Ant. Prop.,* Vol. AP-21, March 1973, pp. 228–229.

[55] Blake, L. V., *Antennas,* Norwood, MA: Artech House, 1984.

[56] Sletten, C. J. (Ed.), *Reflector and Lens Antennas,* Norwood, MA: Artech House, 1988.

[57] LoForti, R., and R. Crescimbeni, "Performance Evaluation of the Imaging Antenna Systems," Final Report Telespazio, ESTEC Contract 5484 83/NL/GM(SC), May 1985.

[58] Rammos, E., A. Roederer, and R. Rogard, "Antenna Technology for Advanced Mobile Communication System," *Proc. First Mobile Satellite Conf.,* JPL, Pasadena, CA, June 1988, pp. 443–449.

[59] Crescimbeni, R., and R. LoForti, "Computer Performance Evaluation of a Multibeam Antenna for Earth Coverage at L-band," *Proc. IEEE AP-S Symp.,* IEEE Cat No. 85 CH2128–7, June 1985, pp. 181–184.

[60] Fitzgerald, W. D., "Limited Electronic Scanning with an Offset Feed Near-Field Gregorian System," Technical Report 486, MIT Lincoln Laboratories, DPC AD-736029, September 1971.

[61] Chen, M. H., and G. N. Tsandoulas, "A Dual-Reflector Optical Feed for Wide-Band Phased Arrays," *IEEE Trans. Antennas Prop.,* Vol. AP-22, No. 4, July 1974, pp. 541–545.

[62] McGahan, R. V., "A Limited-Scan Antenna Comprised of a Microwave Lens and Phased-Array Feed," Report NO. AFCRL-TR-75-0242, Air Force Cambridge Research Laboratories, Massachusetts, May 1975.

[63] Dragone, C., and M. J. Gans, "Imaging Reflector Arrangements to Form a Scanning Beam Using a Small Array," *Bell System Tech. J.,* Vol. 58, No. 2, February 1979.

[64] Nagi, E. C., "Offset Near-Field Gregorian Antenna Scanning Beam Analysis," *RCA Rev.,* Vol. 43, No. 3, September 1982, pp. 504–528.

[65] Woo, K., "Array-Feed Reflector Antenna Design and Applications," *IEE Int. Conf. Ant. Prop. (CAP'81),* York, UK, 1981, pp. 209–213.

[66] Davies, D. E. N., and E. El-Shirbini, "An Array-Fed Parabolic Reflector for Electronic Beam Deflection," *IEE ICAP'83,* Norwich, UK, 1983, pp. 326–330.

[67] Rao, J. B. L., "Bicollimated Near-Field Gregorian Reflector Antenna," Naval Research Lab., Report No. 8658, February 1983.

[68] Lee, B. S., and S. Siddiqi, "Multiple Scanning and Multiple Spot Beam Antenna Design for CONUS Coverage," *Proc. IEEE AP-S Symp.,* 1984.

[69] Hsiao, J. K., "An Electronically Scanning Dual Reflector System," *IEEE AP-S Symp.,* 1984, pp. 471–473.

[70] Pearson R. A., E. El-Shirbini, and M. S. Smith. "Electronic Beam Scanning Using an Array-Fed Dual Offset Reflector Antenna," *IEEE AP-S Symp.,* 1986, pp. 263–266.

[71] Kumar, A., "Study on Hybrid Antennas for Scanning Low Side-Lobe Level and Low Cross-Polarization Level in Millimeter Wave Region," Research Report AK-10, AK Electromagnetique Inc., Quebec, Canada, 1988.

[72] Kumar, A., "An Array Fed Dual Reflector Antenna at 36 GHz," *Proc. IEEE AP-S Symp.,* IEEE Cat No. CH-2654-2/89, June 1989, pp. 1590–1593.

[73] Kumar, A., "Mutual Coupling in a Closely Packed Array," *URSI Symp. Electromagnetics,* Colorado, January 1989.

[74] Albertsen, N. C., S. B. Sorensen, and K. Pontoppidan, ''New Concepts in Multireflector Antenna Analysis, Imaging Systems,'' Final Report S-227-05 ESTEC Contract 5193/82/NL/GM, Rider No. 1, February 1987.

[75] Brueggemann, H. P., *Conic Mirrors,* Focal Press, 1968.

[76] Ackerman, E., S. Wanuga, K. Candela, R. Scotti, W. MacDonald, and J. Gates, ''A 3 to 6 GHz Microwave/Photonic Transceiver for Phased-Array Interconnects,'' *Microwave J.,* Vol. 35, No. 4, April 1992, pp. 60–71.

[77] Adler, R., ''A Study of Locking Phenomena in Oscillator,'' *Proc. IRE Waves Electrons,* Vol. 34, June 1946, pp. 351–357.

[78] Mackey, R. C., ''Injection Locking of Klystron Oscillators,'' *IRE Trans. Microwave Theory Tech.,* Vol. MTT-10, July 1962, pp. 228–235.

[79] Kurokawa, K, ''Injection Locking of Microwave Solid-State Oscillators,'' *Proc. IEEE,* Vol. 61, No. 10, October 1973, pp. 1386–1410.

[80] Carter, P. S., *Microwave Scanning Antennas,* New York: Academic Press, 1966, Ch. 2.

[81] Al-Ani, A. H., and A. L. Cullen, ''A Phase-Locking Method for Beam Steering in Active Array Antennas,'' *IEEE Trans. Microwave Theory Tech.,* Vol. MTT-22, No. 6, June 1974, pp. 698–703.

[82] Stephan, K. D., ''Inter-Injection-Locked Oscillators for Power Combining and Phased Arrays,'' *IEEE Trans. Microwave Theory Tech.,* Vol. MTT-34, No. 10, October 1986, pp. 1017–1025.

[83] Navarro, J. A., and K. Chang, ''Electronic Beam Steering of Active Antenna Arrays,'' *Electron. Lett.,* Vol. 29, No. 3, February 1993, pp. 302–304.

Chapter 4
Acousto-Optics Time Delays for Phased Array Antennas

4.1 INTRODUCTION

The technology of optical signal processing is used to steer a broadband phased array antenna system. The parallel nature of light, along with the high-packing density available with integrated optical components, allow a parallel processing approach to be taken for some of the rather sophisticated alogrithms necessary for applications such us radar/antenna signal processing. In the case of electronic beam steering including nonfiber optical signal processor for antenna beamforming network, the performance of the system is narrowband. Therefore, beam-pointing (squint) error in the phased array antenna system is most common. The conventional diode phase shifters are sequentially controlled by a numerical computer, which makes the steering relatively slow. The weight and size of the hardware are concentrated at the antenna back-plane. There are problems in designing the power divider that drives all the elements from a single source. These problems can be illuminated completely by using the true time-delay beamforming using acousto-optics. This chapter describes a short history (background) of the acousto-optics, generation, and control of the frequency-dependent phase shift and a description of the acousto-optics delay line. Applications of the acousto-optics delay line for the receiving and transmitting phased array antennas have been described. Theoretical and experimental results are presented for the one-dimensional and two-dimensional beam scanning of linear and planar phased arrays.

4.2 BACKGROUND OF THE ACOUSTO-OPTICS

The invention of acousto-optics with the interaction of sound and light was predicted by Brillouin [1] in 1922. Lucas et al. [2] and Debye et al. [3] experimentally verified the interaction of sound and light in 1932. Debye–Sears diffraction is analyzed by using a moving phase-grating model as was first done by Raman and Nath [4]. Nath [5] predicted that a sound cell can be used to modulate light by using the Raman-Nath mode. During the discussion on the Bragg mode [6], he stated that the Bragg mode can also be used for the modulation of light by a sound cell. For the Raman-Nath type diffraction, the acousto-optics parameter $Q = (2\pi\lambda_1 L)/(n/\Lambda)$ is less than or equal to 0.3. The symbols λ_1 and Λ represent, respectively, the wavelengths of the light wave (in free space) and the acoustic wave, n is the effective refractive index of the medium, and L denotes the aperture of the acoustic wave. In the Bragg diffraction, the value of Q is larger than or equal to 4π. The essential properties of Bragg diffraction has been explained in many different ways by Quate et al. [7], Gordon [8], Adler [9], King et al. [10], Whiteman et al. [11], and Korpel [12–15]. Approximately 50 years after Brillouin's analysis, all the phenomena he predicted had been put to use. Brillouin predicted a basic Doppler shift equal to the sound frequency. The phenomenon of frequency shifting became important only in recent times and forms, for example, as the basis of heterodyning techniques in modern signal processing applications.

We shall limit our discussion to the Bragg-type diffraction in this chapter because this type of diffraction is capable of larger modulation bandwidth and dynamic range.

4.3 DELAY LINE CONCEPT

RF/microwave phase shift using optical processing is known as optical heterodyning. In the past, the optical heterodyning method has been used for digital transmission of TV signals with fiber optics. Okoshi [16] reviewed the recent developments on a heterodyne system for optical fiber communications. Saito et al. [17] reported optical hetrodyne detection experiments using a semiconductor laser transmitter and a local oscillator. Bachus et al. [18] used a heterodyne-type television transmission with a fiber optic.

The optical heterodyne technique for achieving RF/microwave phase shift using optical processing is shown in Figure 4.1. In the figure, the laser output has been split into two paths using an optical splitter. In one path, the optical signal frequency is shifted by an amount f_0 (the microwave signal frequency) and then optically phase shifted by an amount Φ_0. The phase and frequency shifted beams are summed with the light from the other path, which provides a phase reference. These two collinear beams are incident on a photodetector that responds only to the time-averaged intensity of the incident light. The resulting electrical signal at the output of the photodetector diode provides the time-varying term $\cos(2\pi f_0 t - \Phi_0)$. The resulting beat frequency term is phase shifted by the same value Φ_0 as the optical beam. The optical heterodyne process serves as an optical-to-RF/microwave phase converter, and it is easy to achieve an RF/microwave phase shift up to 2π.

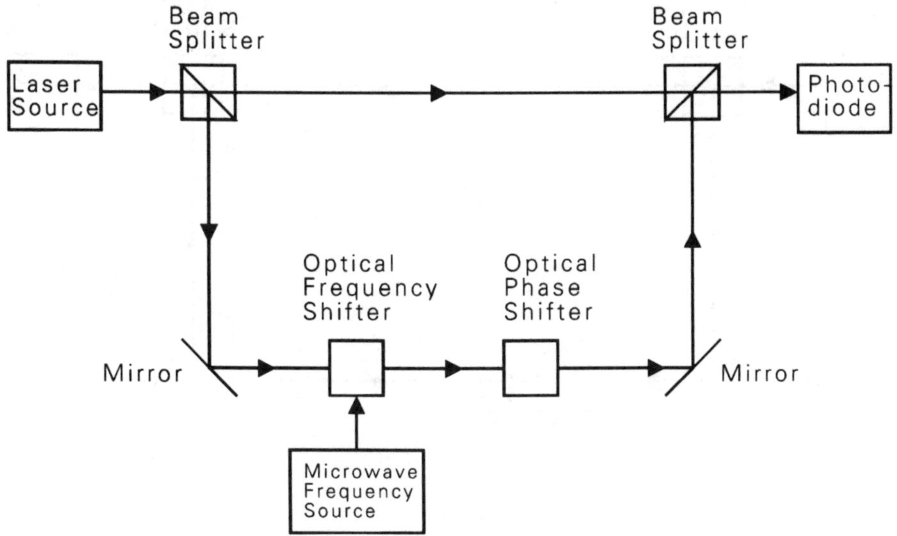

Figure 4.1 Optical delay line concept.

4.4 THEORY OF THE DELAY LINE

Many researchers [19–35] have described an acoustic-optic-based signal processor to achieve a broadband beam-steering performance for a phased array antenna. They have also reported a method for obtaining substantial microwave phase shift using an optical heterodyning (photomixing) process as described in Section 4.3. Korpel [15] described a delay line theory by utilizing an acoustic-optic (Raman-Nath mode or Bragg mode) cell as the frequency shifter in a heterodyne configuration to achieve continuous variable time delay.

In 1982, an optical implementation of the frequency-shifting Bragg cell was reported by Kumar [19] as shown in Figure 4.2(a). In the figure, the laser beam is divided into two sections through an optical splitter. In the top section the beam frequency is f, and in the bottom section the reflected beam from the mirror is modulated with a RF/microwave signal (frequncy f_0) in the Bragg cell. The frequency of the modulated signal is $f + f_0$, which meets with the signal of the top section of an optical combiner. The time delay between the two beams of frequencies f and f_0 is τ. Toughlian et al. [24] reported a delay line system that is shown in Figure 4.2(b). It is seen that a frequency-modulated laser beam incident on a grating will have its various frequency components spatially dispersed as shown. A cylindrical lens is used to stop the spread and image the spatially separated frequency components on a reference plane in front of a mirror capable of both translation and rotation. Each frequency component experiences a different delay due to the different

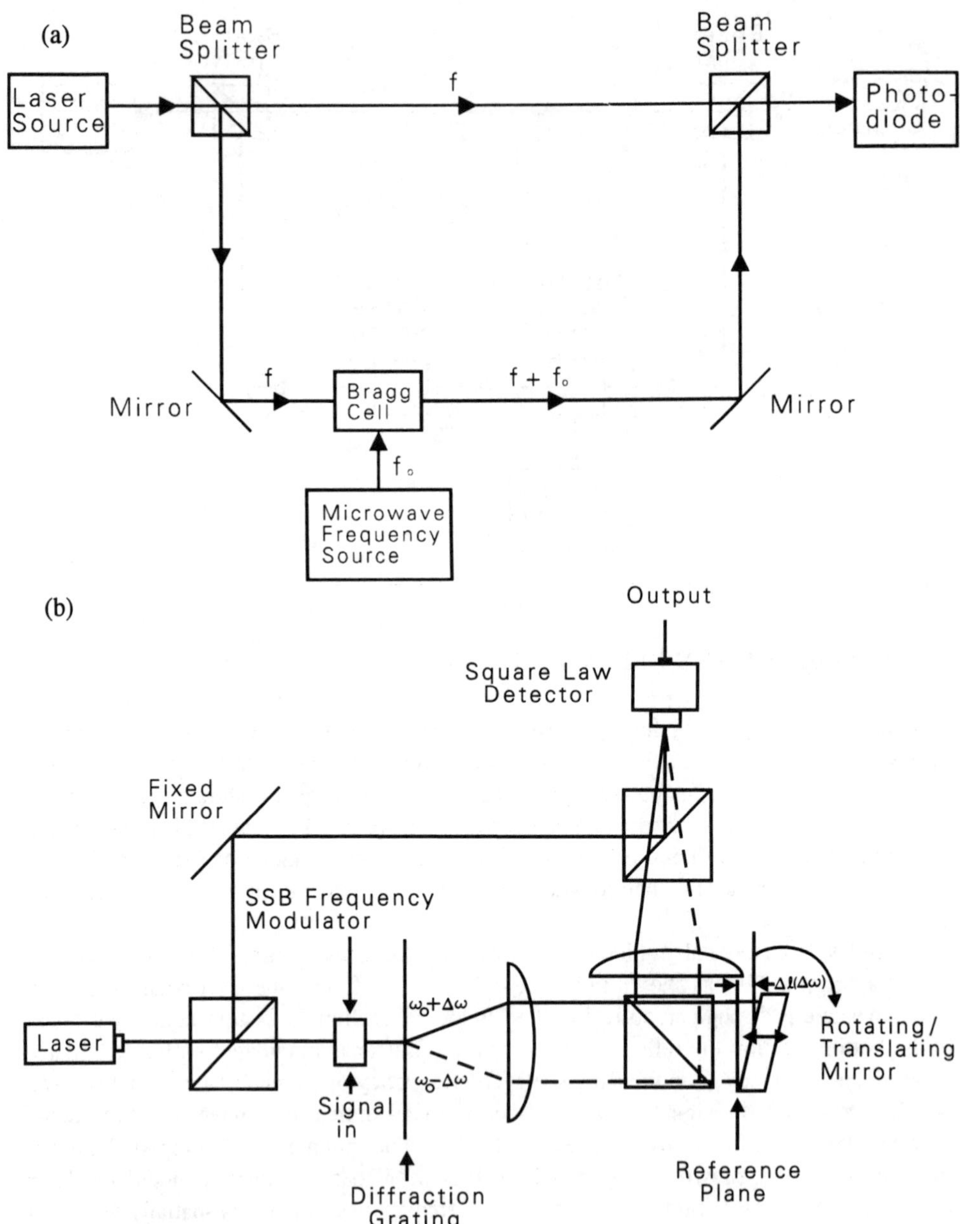

Figure 4.2 (a) Delay line using a Bragg cell [19]. (b) Frequency-compensated heterodyne scheme. (© 1990 IEEE.)

optical paths traveled. The path difference can be varied by rotation and translation of the mirror. Since the heterodyne process preserves phase, an RF/microwave phase shift equal to the optical phase shift results. This RF/microwave phase shift varies linearly with frequency and in effect provides true time delay. Quantitatively, the differential time delay is given by

$$\Delta t = \beta \Delta l(\omega_M)/\omega_M \approx \omega_0 \Delta l(\omega_M)/\omega_M c \tag{4.1a}$$

where, β is the propagation constant for optical angular frequency ω_0, RF/microwave angular frequency ω_M, $\beta = (\omega_0 + \omega_M)/c$, Δl is the differential optical path length, and c is the velocity of light in the medium. It is shown in Figure 4.2(b) that the differential path length Δl is a function of RF/microwave frequency, which yields a true time delay independent of RF/microwave frequency as desired.

In 1992, Zmuda et al. [28] used a simplified system. For this simplified system, a straightforward geometrical optical analysis shows that the time delay can be calculated from

$$T = \tan(2\Theta)f/v \tag{4.1b}$$

where Θ is the mirror tilt angle, f is the lens focal length, and v is the acoustic velocity in the Bragg cell.

In this section, we show how the time delay is determined by the orientation of the optical reference beam at the photodiode. Figure 4.3(a) shows a practical delay line system. In the figure, the laser output of the delay line system is split into a plane wave local oscillator and acts as a phase reference. The first split signal is sent to an acoustic-optic cell (Bragg cell) modulator where an RF/microwave signal is applied and modulation is obtained with the optical signal. This modulated signal travels through a *Fourier transform* (FT) lense and the photodiode. The second split signal is reflected through a fixed mirror and a Bragg cell where RF antenna steering is modulated. The modulated steering signal is reflected through a mirror and incident on the photodiode with the signal from the first section. The output resultant signal from the photodetector is the delay line signal. The optical reference beam traveling as the second split signal in Figure 4.3(a) is a plane wave at the photodetector diode, making an angle Θ with the face of photodetector diode. Figure 4.3(b) shows the reference plane wave and signal spectrum on the photodetector diode. The reference plane wave at the photodetector diode is given by [27]

$$r(t,y) = e^{j2\pi\,(f_0 t - (y/\lambda)\sin\theta)} = e^{j2\pi f_0 (t - (y/c)\sin\theta)} \tag{4.2}$$

where t is time, y is the physical distance along the photodetector diode, λ is the optical wavelength of the reference beam, f_0 is the optical carrier frequency, and c is the velocity of light. In Figure 4.3(a), the first split optical signal is modulated by the spectrum of the

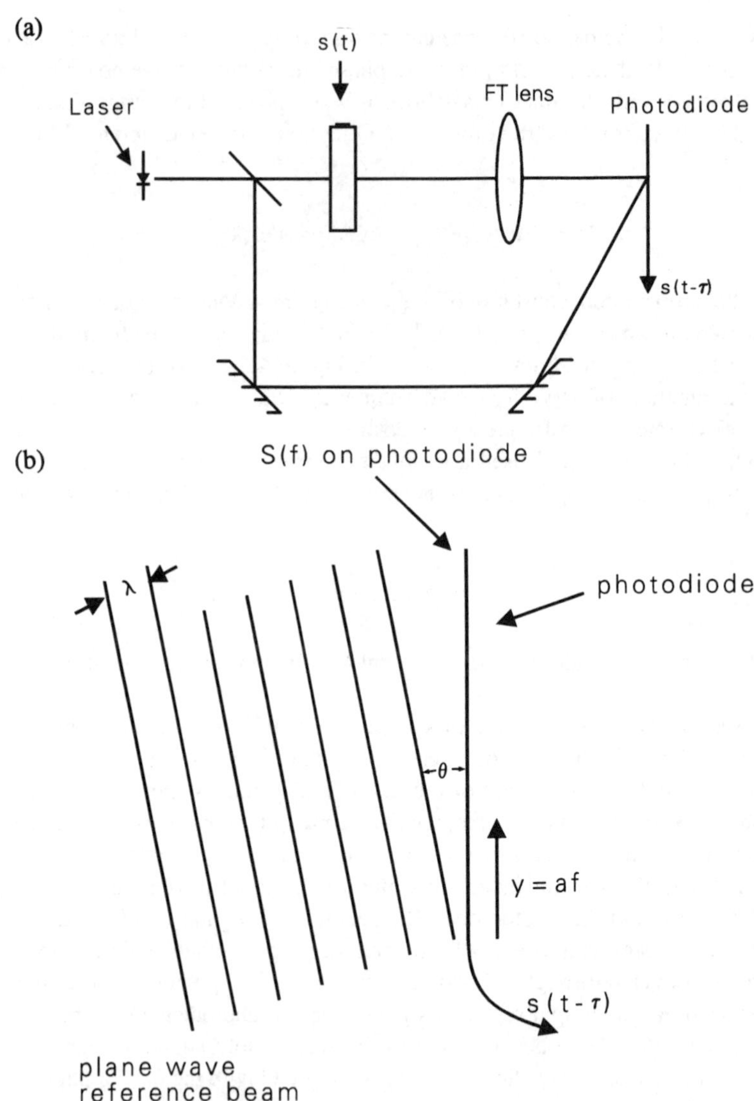

Figure 4.3 (a) A practical delay line system, and (b) reference plane wave and signal spectrum on the photodiode. (© 1992 SPIE.)

RF/microwave signal *s(t)* and by tones oscillating the spectral frequencies of *s(t)* also illuminates the photodetector diode. The modulated optical beam has the following functional form at photodiode:

$$S(f) \cdot e^{j2\pi ft} e^{j2\pi f_0 t} \tag{4.3}$$

The variation of the RF/microwave signal incident to the acousto-optic Bragg cell is given by

$$e^{j2\pi ft} \tag{4.4}$$

The phase of the reference plane wave on the photodiode varies linearly with position as indicated in (4.2). Since spectral frequency f in the Fourier transform at the face of the photodiode is proportional to the physical distance y along the photodiode, a time delay τ can be calculated by

$$\tau = -\frac{y}{f}\frac{f_0}{c}\sin\theta \tag{4.5}$$

In the above equation, τ is proportional to the sine of the angle between the reference plane and the photodiode. The reference wave at the photodiode of (4.2) can be expressed as

$$r(t,f) = e^{j2\pi f\tau}e^{j2\pi f_0 t} \tag{4.6}$$

The detection of signals by the photodetector diode results in multiplication of the signal spectrum by the reference plane wave, and each spectral component is phase shifted by an amount proportional to the frequency of the component that provides a time delay. The output $d(t)$ of the photodetector diode is the square-law detection at each instant of time and is integrated with respect to frequency f as

$$d(t) = \int |r(t,f) + S(f)\cdot e^{j2\pi ft}e^{j2\pi f_0 t}|^2 df \tag{4.7}$$

The value of $r(t,f)$ can be introduced from (4.6) to (4.7) and we have

$$d(t) = \int |e^{j2\pi f\tau}e^{j2\pi f_0 t} + S(f)\cdot e^{j2\pi ft}e^{j2\pi f_0 t}|^2\, df \tag{4.8}$$

which can be simplified to

$$d(t) = \text{bias} + 2\text{Re}\left[\int S(f)e^{j2\pi f(t-\tau)}\, df\right] \tag{4.9}$$

where $\text{Re}[\int S(f)e^{j2\pi f(t-\tau)}\, df]$ denotes the real part of the complex quantity and the photodetector diode output is simplified to

$$d(t) = \text{bias} + 2\text{Re}[s(t-\tau)] \tag{4.10}$$

The input signal $s(t)$ that drives the Bragg cell is real, and the bias allows negative and positive values of the delayed signal to be represented as shown in (4.10)

4.5 TIME DELAYS ARRANGEMENT FOR PHASED ARRAY ANTENNAS

There is a requirement of one photodiode for each element of the array to create a delayed replica of the signal to be transmitted. Each of these photodiodes must be illuminated by a plane wave with a different angle of incident in order to generate the set of time delays required to steer the array beam. To scan the formed beam, the changes in the time delays must remain proportional to one another, with the constants of proportionality determined by the separation between the array elements. We use movable mirrors or multichannel Bragg cells to control the orientations of the reference plane waves and maintain the proportionality relations between the changes of time delay.

4.5.1 Time Delay Arrangement for Transmit Array Antennas

Gesell et al. [26, 27] described the acousto-optic architecture for generating and controlling a set of time delays for linear transmitting arrays as shown in Figure 4.4 In the figure, the signal is produced from a laser diode and splits into two sections. In the first section, coherent light from the laser source passing through the Bragg cell is modulated by the signal *s(t)*. This modulated light passes through a Fourier transform lens, which produces a spectrum of the signal, and the signal illuminates a column of wideband photodiodes.

The split signal in the second section in Figure 4.4 is reflected from the mirror and incident on the Bragg cell. An electrical steering signal (frequency, f) drives the Bragg cell in the beam steering leg of the interferometer. The acoustic velocity of the signal in the

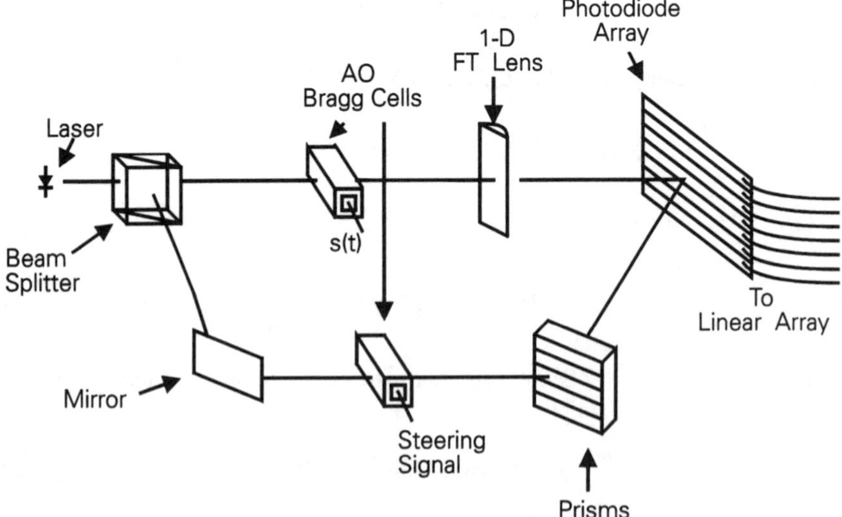

Figure 4.4 Time delays for a transmitting array. (© 1992 SPIE.)

Bragg cell is v_a. The output signal from the Bragg cell is deflected from the optical reference beam by an angle of ζ and is denoted by

$$\zeta = 2 \cdot \arcsin\left[\frac{\lambda}{v_a} \cdot f\right] \tag{4.11}$$

The angle of incident of the deflected plane wave makes an angle ω on the face of the prism. The index of refraction of the prism and air are n_1, n_2, respectively. The angle ω is measured from the normal to the face of the prism and given by

$$\omega = \zeta + \delta \tag{4.12}$$

where δ is the orientation angle of the prism measured between the normal to the face of the prism and the undeflected optical reference signal. The output signal leaves the face of the prism at angle ν from the normal to the output face and it is calculated by

$$\nu = \arcsin\left[\frac{n_2}{n_1} \cdot \sin\left[-\arcsin\left[\frac{n_1}{n_2} \cdot \sin(\omega)\right]\right] - \gamma\right] \tag{4.13}$$

In Figure 4.5, the stack of prisms provides a single-channel Bragg cell to steer the array, since a prism amplifies the changes of the angle of deflection out of the Bragg cell, with the angle amplification determined by the wedge angle (γ) of the prism. Therefore, with a stack of prisms it should be possible to maintain the required proportionality relations between the changes in the relative time delays. However, the set of time delays is dictated by the frequency of the single steering signal and by the stack of prisms in the beam-steering section.

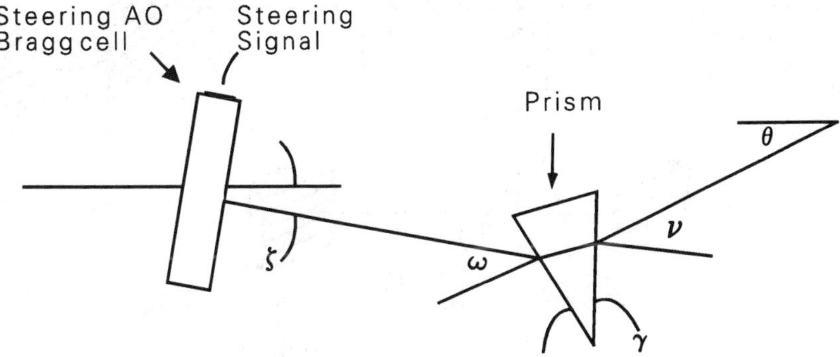

Figure 4.5 Bragg cell/prism optical beam control. (© 1992 SPIE.)

The output optical signal from the prism is incident on the photodiode, which makes an angle θ from the normal to the photodiode that is given as

$$\theta = \nu - (\gamma - \delta) \tag{4.14}$$

The wedge angle of the prism is γ, and angles for describing signal directions from Bragg cell to the input of photodiode are shown in Figure 4.5.

Signals traveling via both sections (first and second) illuminate photodiodes, and output microwave signals are fed to the antenna elements via microwave amplifiers as shown in the figure. The direction of propagation of a beam formed by the linear array is given by

$$\alpha = \arcsin\left(\frac{c \cdot \tau}{b}\right) \tag{4.15}$$

4.5.2 Time Delay Arrangement for Receiving Array Antennas

Figure 4.6 shows a schematic to generate time delays for a receiving one-dimensional linear array. Gesell et al. [27] mentioned that the outputs of the array elements must be delayed relative to one another and then summed to form a beam in a particular direction with the receiving array. The relative delay between array elements is determined by the

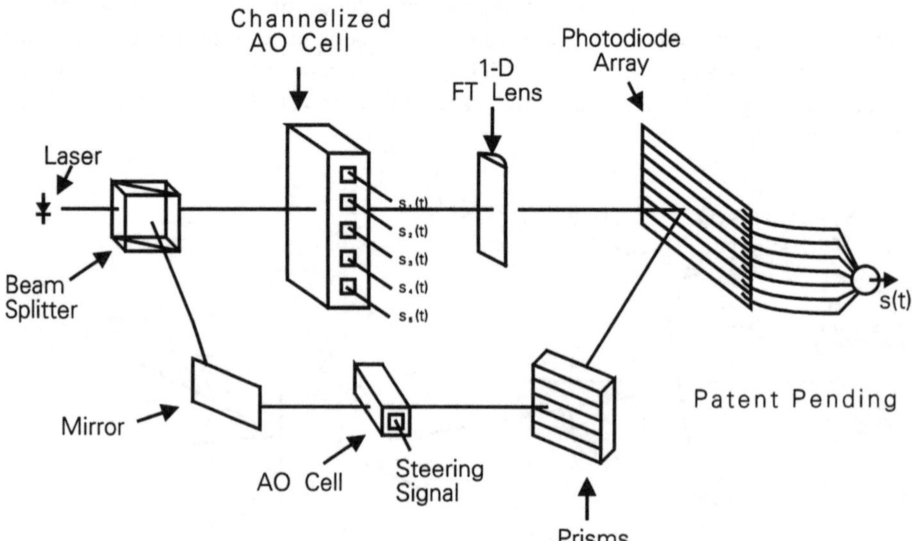

Figure 4.6 Time delays for a receiving array. (© 1992 SPIE.)

spacing between the array elements and the angle between the beam and the normal to the array. The principle of the receiving array is the same as the transmitting array. However, in the case of the receiving array, the number of channels in the Bragg cell are the same as the number of linear array elements and, therefore, a signal from each element drives a channel of the multichannel Bragg cell. The received signal $s_l^{(t)}$ to $s_j^{(t)}$ corresponds to the elements $l, \ldots, j$ of the linear array. The steering signal is derived from the Bragg cell as shown in the figure.

4.5.3 Computed Results for Transmit and Receive Array Antennas

Gesell et al. [26, 27] used four antenna array elements positioned at 0.0m, 0.2m, 0.4m, and 2.0m; 1-GHz signal bandwidth; 2.9-mm width of the photodiode; and slow shear mode TeO for the beam-steering Bragg cell. A plot of the antenna beam scan angle versus steering control frequency without prisms for the two end elements of the array is shown in Figure 4.7(a). The plot shows that changing the frequency of the steering signal by 250 kHz scans the beam 20° away from the array normal.

A plot of the beam scan angle versus the control frequency is shown in Figure 4.7(b) for a single steering Bragg cell and a stack of prisms. The antenna element pair spacings are 0.2m, 0.4m, and 2.0m. The wedge angle (γ) of the prism and orientations are optimized for scanning the array from 0° to 60°. It is shown in the figure that a change 10 MHz in the steering control frequency scans the beam by 60°.

Figure 4.7(c) shows plots of the difference in directions of the beams formed by the individual antenna element pairs when the single-channel Bragg cell and a stack of prisms are used. The figure shows that the maximum difference in the direction of propagation between beams formed by the pair of elements separated by 0.2m and 2m is 0.14°, and the maximum difference in the direction of propagation between beams formed by the pairs of elements separated by 0.2m and 0.4m is 0.5°.

4.6 EXPERIMENTAL RESULTS ON DELAY LINES AND PHASED ARRAY ANTENNAS

Toughlin et al. [24] and Zmuda et al. [28] demonstrated the delay line performance experimentally by arranging a setup as shown in Figure 4.8. They performed the measurements with a 5-mW HeNe laser beam operating at a wavelength of 632.8 nm. An acousto-optic modulator is used and is driven with the source of the network analyzer. The modulator provides the spatial frequency dispersion and frequency shift in the laser by 70 MHz with a 25-MHz modulation bandwidth. A cylindrical lens of 15-cm diameter is used to stop the angular spread and image the resulting line onto the phase compensating mirror. The mirror is mounted on two computer-driven motorized stages capable of precise translation within 25 nm and rotation within 0.175-mrd resolution (this motion is used to obtain frequency-dependent phase compensation). The light reflected from the compensating mirror is summed with a portion of the laser output and mixed in the detector. The delay line phase performance as a function of frequency is measured with a two-port network analyzer. Figures 4.9 and 4.10 show the electrical phase (Φ) response of the system as a function of RF frequency *f*. Five curves are shown in Figure 4.9, each of which represents a different tilt of the mirror, that is, a different electrical delay. The time delay in seconds is calculated by the slope of the phase versus frequency curve, that

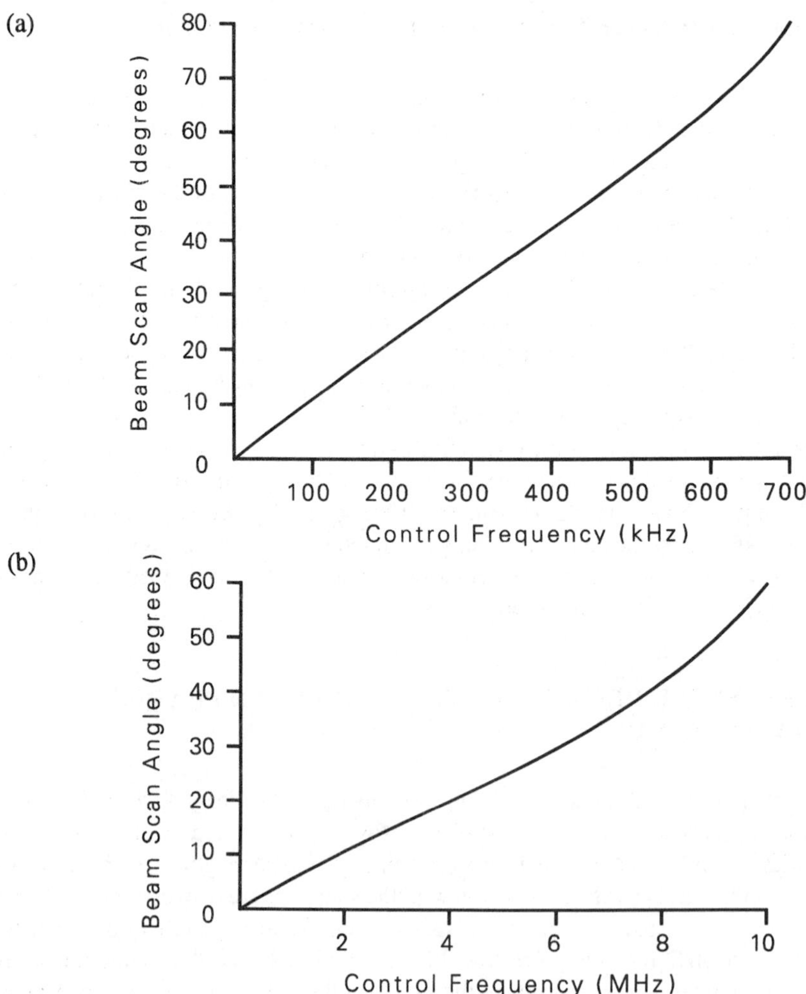

Figure 4.7 Beam scan angle versus steering control frequency (a) without prism and (b) with prism. (c) Deviation in the angle of propagation from different pairs. (© 1992 SPIE.)

(c)

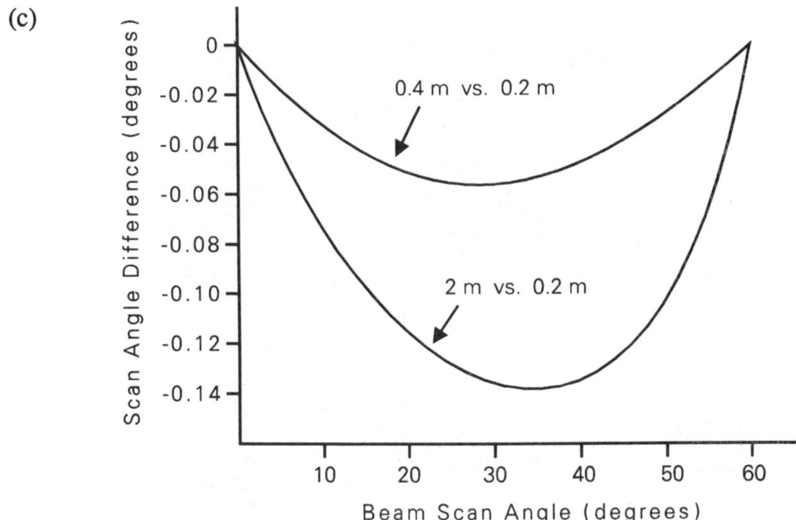

Figure 4.7 (continued)

is, $\Delta\Phi/\Delta\ (2\ \pi\ f)$. Larger delay is shown in Figure 4.10, when the tilt angle of the mirror increases.

A basic system of steering a phased array antenna is explained in Chapter 3. A compact system that incorporates the multiple delay lines that are necessary for phased array antennas is presented in the spatially integrated optic implementation of Figure 4.11. Such an arrangement provides mechanical stability, eliminates the drift problems encountered in free-space optical systems, and provides the ability to achieve higher packing densities. The basis of this integrated optical implementation is known as a *segmented mirror device* (SMD). This type of device can be utilized to provide the phase shifts that are necessary to achieve a true-delay radiation pattern of a phased array antenna. The array factor of a linear array of isotropic point sources spaced a distance d apart can be given by the well-known equation:

$$F_1 = \sum_m A_m \exp\ \{j[\varphi + md \sin \theta(\omega_c + \Delta\omega)/c]\} \tag{4.16}$$

where ω_c, is the RF carrier frequency, $\Delta\omega$ is the modulating RF/microwave frequency, θ is the given observation angle, φ is the required RF phase shift for the mth array element, and A_m is the amplitude of the mth element.

In this chapter, we have discussed that the phase shift can be obtained via an optical heterodyne process. In this circumstance, the scan angle of phased array antennas can be given by

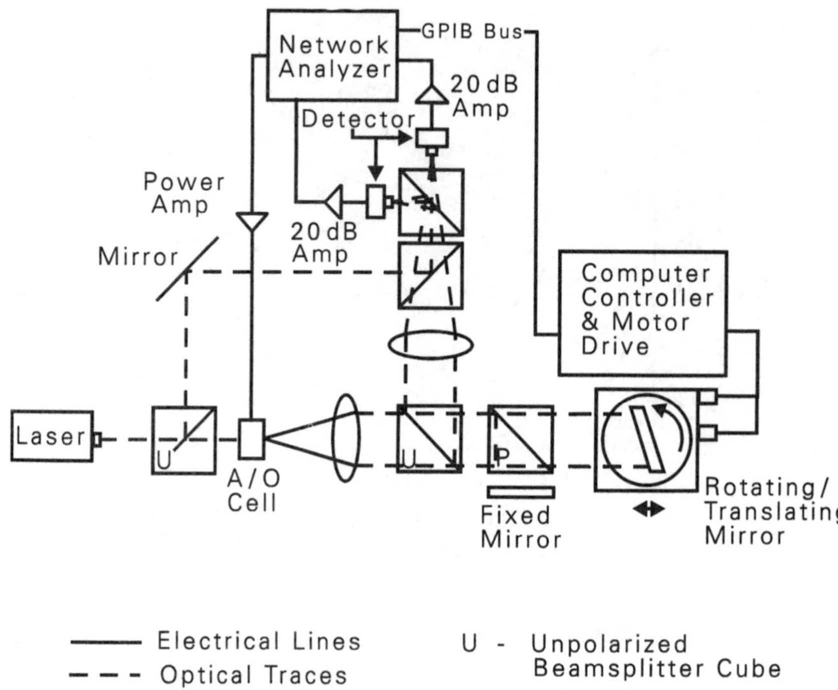

Figure 4.8 Experimental setup for a frequency-compensated heterodyne system. (© 1990 IEEE.)

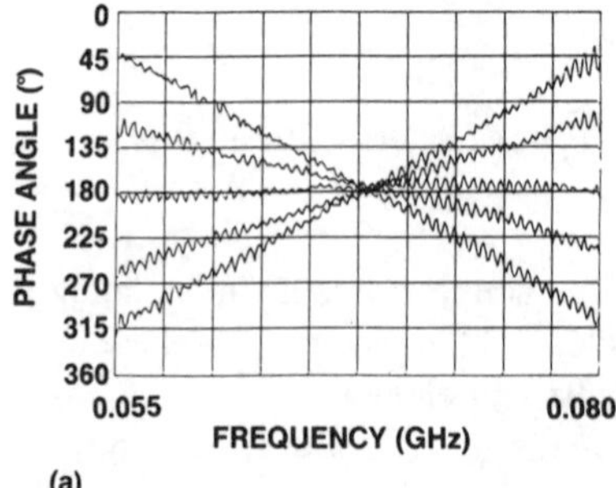

Figure 4.9 Measured phase versus frequency. (© 1990 IEEE.)

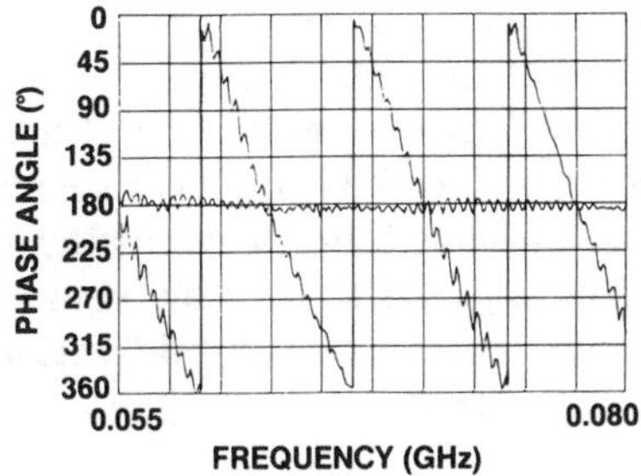

Figure 4.10 Phase versus frequency for larger delays. (© 1990 IEEE.)

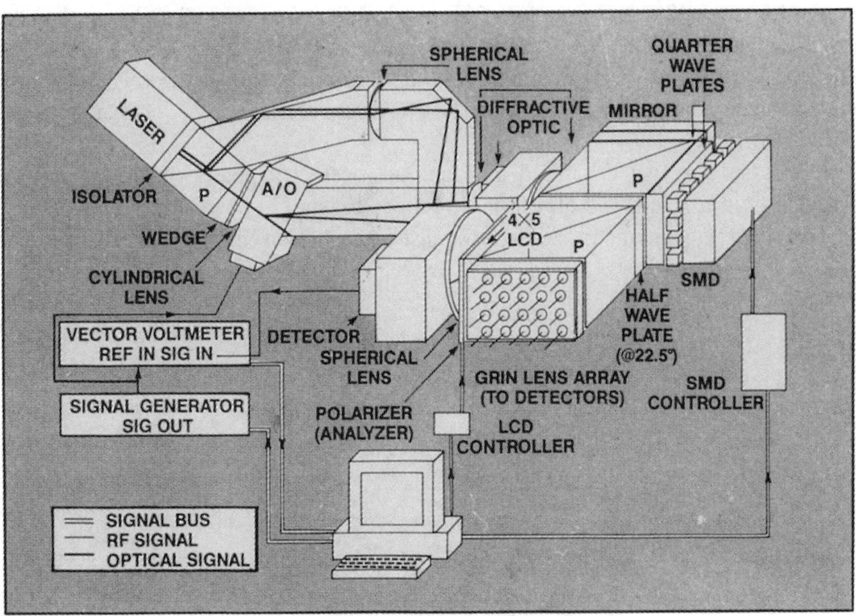

Figure 4.11 A 20-element photonic/microwave phased array [28].

$$\theta_0 = \text{Arc sin } [\{-md(\omega_c + \Delta\omega)\}/(\Delta l\ \omega_0)] \tag{4.17}$$

where Δl is the differential path length and ω_0 is the optical angular frequency.

A set of 20 delay lines is integrated with a 20-element broadband phased array antenna as shown in Figure 4.11. The output of the laser beam is split into two orthogonally polarized beams, including a local oscillator beam and a beam that is frequency shifted by the Bragg cell. These two beams are combined by a polarizing cube and sent to a diffractive optic device, by which the phase reference beam and all the delay information contained in the Bragg cell are replicated twenty times. These twenty beams are sent to a polarizing cube in order to separate the local oscillator beams from the frequency-shifted beams. The signal beams are imaged into a 20-element array of tiltable mirrors. To achieve a proper polarization relation between the optical signal and local oscillator beams, some additional optics are used to steer the light beam at both outputs. The detected output of the lens array provides delay in RF/microwave signals, which is used to steer the beam of the antenna array.

4.7 TWO-DIMENSIONAL ARRAY ANTENNA BEAM STEERING

In Section 4.5, we described the system that provides beam-scanning capability along one direction. Lin et al. [29] and Riza [30–33] reported two-dimensional (2D) antenna beam-steering techniques. In these techniques, the processor uses acousto-optic devices in a crossed Bragg cell configuration and the two-dimensional beam position is controlled via two independent analog control signals. The phase and carrier can be independently set to a desired value.

In this section, we described a technique to steer a phased array antenna beam independently in azimuth and elevation directions by simply controlling the phase of the signals driving the elements in the two-dimensional array. In Figure 4.12, x' and y' are the

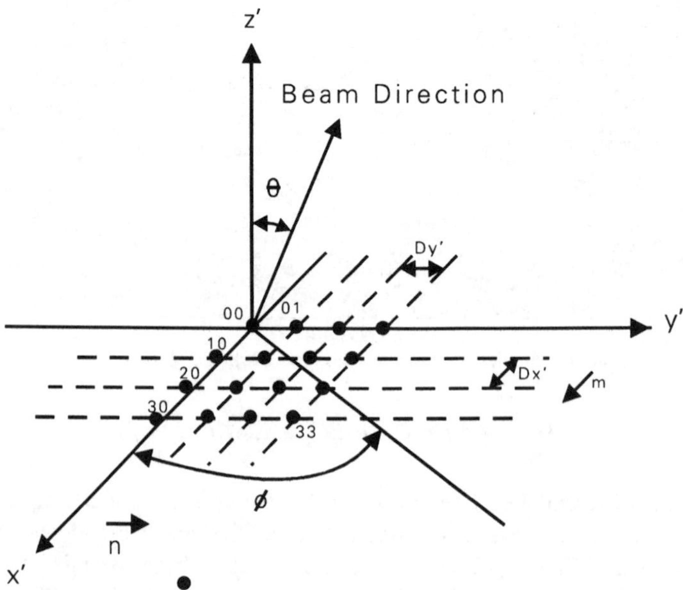

Figure 4.12 Planar phased array elements and beam-scanning notation. (© 1992 SPIE.)

beam-space cartesian coordinates, where θ is the angle of scan measured from the broadside direction and ϕ is the plane of scan measured from the x'-axis. A simplified technique for describing the two-dimensional beam position in terms of direction cosine is proposed by Von Aulock [34]. For a particular desired beam position corresponding to the direction cosines cos $(\alpha_{x's})$ and cos $(\alpha_{y's})$ for a linear phase taper,

$$\cos(\alpha_{x'}) = \sin(\theta)\cos(\phi) \tag{4.18a}$$

$$\cos(\alpha_{y'}) = \sin(\theta)\sin(\phi) \tag{4.18b}$$

The signal in the mnth element in the array located at (mDx', nDy') is given by Von Aulock [34] as

$$i_{mn}(t) = A_{mn}\cos[\omega_r t + m\phi_{x'} + n\phi_{y'}) \tag{4.19}$$

where A_{mn} is the mnth element signal amplitude, ω_r is the element carrier frequency, $\phi_{x'}$ is the element-to-element phase shift along the x'-direction, $\phi_{y'}$ is the element-to-element phase shift along the y'-direction, $\phi_{x'} = (2\pi\lambda_r)\, D_{x'} \cos(\alpha_{x's})$, $\phi_{y'} = (2\pi\lambda_r)\, D_{y'} \cos(\alpha_{y's})$

$D_{x'}$ is the antenna inner-element spacing along the x'-direction, $D_{y'}$ is the antenna inner-element spacing along the y'-direction, and λ_r is the antenna carrier wavelength.

The antenna beam position in the spherical coordinate is given by

$$\phi_s = \tan^{-1}[\cos(\alpha_{y's})/\cos(\alpha_{x's})] \tag{4.20a}$$

$$\theta_s = \sin^{-1}[\cos(\alpha_{x's})/\cos(\phi_s)] \tag{4.20b}$$

Acousto-optics system for the two-dimensional phased array system.

Riza [33] presented an acousto-optics system that can generate the appropriate phase signals that provide the two-dimensional beam steering in a planar phased array antenna. He used four one-dimensional acousto-optics devices (Bragg cells) in a crossed cell, a two-dimensional photo detector diode array, imaging and Fourier transforming lense, and three signal generators to provide the antenna element carrier and azimuth/elevation control signals as shown in Figures 4.13(a) and 4.13(b). In this system, the laser light is collimated along the x-direction (lens S_1) and focused along the y-direction by a cylindrical lens C_1. The light from lens C_1 is incident to the Bragg cell (AOD1), and undiffracted and diffracted beams are spatially separated by the spherical lens S_1, which forms two separate vertical slits of light. The RF/microwave signal fed in AOD1 is

$$S_{\text{AOD1}} = a\cos(\omega_c + \omega_1)t \tag{4.21}$$

where a is the amplitude of the RF signal, ω_c is the RF/microwave angular frequency (antenna carrier angular frequency), and ω_1 is the RF antenna scan control angular fre-

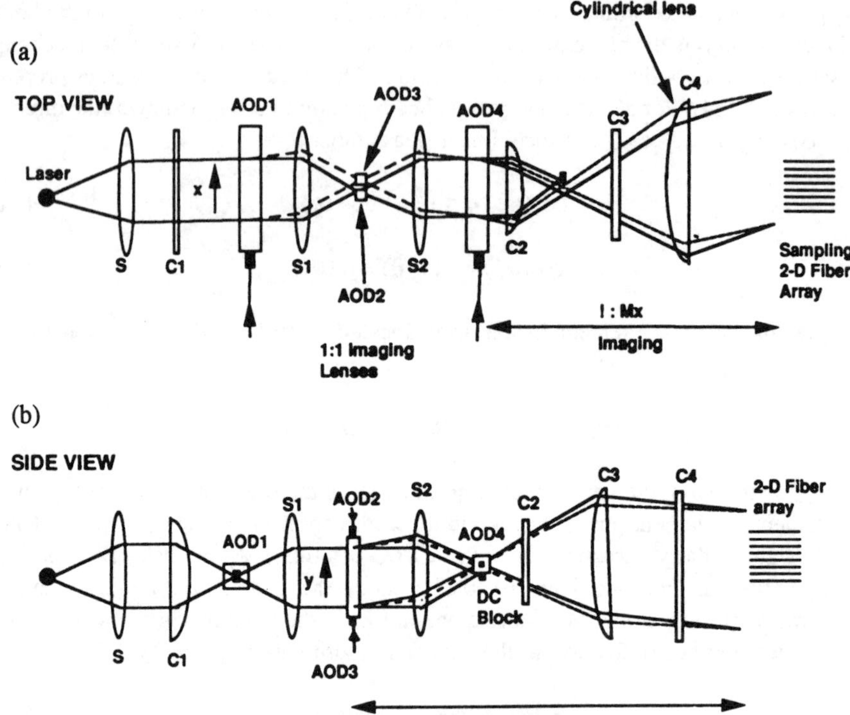

Figure 4.13 Two-dimensional phased array antenna scanning using acousto-optics: (a) top view and (b) side view. (© 1992 SPIE.)

quency that controls the antenna array beam scanning in one direction. Bragg cells AOD2 and AOD3 are orthogonally oriented along the *y*-direction, and they are positioned such that the undiffracted and diffracted slits of light from AOD1 fall within AOD2 and AOD3 acoustic columns, respectively. The RF/microwave signals fed in AOD2 and AOD3 are

$$S_{\text{AOD2}} = a\cos(\omega_c - \omega_2)t \tag{4.22a}$$

$$S_{\text{AOD3}} = a\cos(\omega_c + \omega_2)t \tag{4.22b}$$

where ω_2 is the RF antenna control frequency that controls the beam other direction or orthogonal direction to θ_s, ϕ_s. The spherical lens S_2 focuses the diffracted and undiffracted beams from AOD2 and AOD3 into vertically separated horizontal slits. However, the acoustic column of AOD4 is placed such that it encloses the horizontal slit from the diffracted rays from AOD2 and AOD3. The other horizontal slit from the undiffracted light from AOD2 and AOD3 is blocked. The undiffracted light from AOD3 is blocked in the plane of AOD4. The diffracted rays from AOD3 passes through AOD4. The RF/ microwave signals feeding the AOD4 is

$$S_{\text{AOD4}} = a\cos(\omega_c - \omega_1)t \tag{4.23}$$

The beams through AOD4 are interfered on a two-dimensional photodetector diode array, which in turn generates the necessary signals for two-dimensional steering for a planar phased array. Then signals are magnified M_x times along the x-direction and M_y times in the y-direction. The current generated from the mnth photodetector diode pair in the sampling two-dimensional array can be approximated similar to (4.19) and given by [33]

$$i_{mn}(t) = G\cos[\omega_r t - m\psi_x - n\psi_y] \tag{4.24}$$

where $\omega_r = 4\omega_c$, G is a constant,

$$\Psi_x = \frac{2\omega_1 l_x}{M_x v_a} \tag{4.25}$$

$$\Psi_y = \frac{2\omega_2 l_y}{M_y v_a} \tag{4.26}$$

In the above equations l_x and l_y are the interfiber spacing along the x- and y-directions, $m = 0, 1, 2, \ldots, M$, and $n = 0, 1, 2, \ldots, N$. The two-dimensional beam-steering conditions for the optical processor for positioning the antenna beam at the direction cosine $\cos(\alpha_{x's})$ and $\cos(\alpha_{y's})$ are given by

$$\psi_x = \phi_{x'}, + p2\pi = 2\pi(D_{x'}/\lambda_r)\cos(\alpha_{x's}) + p2\pi \tag{4.27}$$

$$\psi_y = \phi_{y'} + q2\pi = 2\pi(D_{y'}/\lambda_r)\cos(\alpha_{y's}) + q2\pi \tag{4.28}$$

where $p, q = 0, +1, +2, \ldots$

Riza [33] made the experimental setup in the laboratory, and apparatus are shown in Figures 4.13 (a,b) and 4.14. Table 4.1 provides the list of equipment used for this experimental work.

Riza [33] used the two-dimensional photodetector diode array that is simulated by two avalanche photodiodes (Hamamatsu model S2381 operated with 159-V bias) that are placed with inner-detector spacings of $l_x = 6$ mm and $l_y = 5.6$ mm. The Bragg cells AOD1 and AOD4 are driven by 20.3V peak-to-peak at 50 Ω and 70-MHz RF signals. AOD2 and AOD4 are driven by 22.7V peak-to-peak at 50 Ω and 75-MHz RF signals. Figures 4.15 and 4.16 show the measured phase shift versus control frequencies for alignment of the x- and y-directions, respectively. This demonstration proves that a desired phase shift can be achieved along the two independent directions (x and y) so as to steer a planar phased array antenna beam in the azimuth and elevation directions.

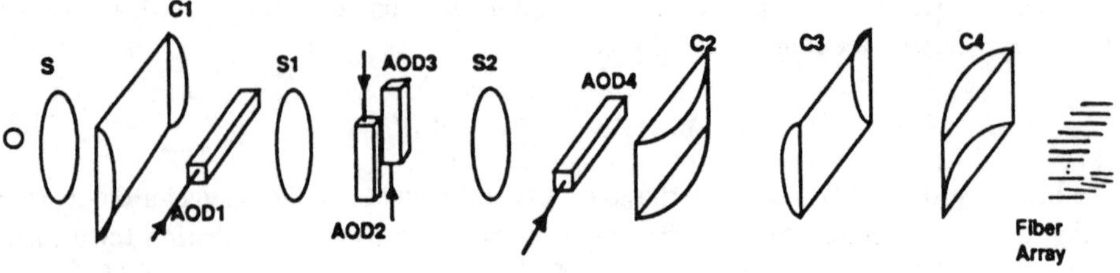

Figure 4.14 Two-dimensional scan control architecture. (© 1992 IEEE.)

Table 4.1
List of Components Used for the Experiment by Riza

Components	Specifications
1. Ar+ ion laser	= 514 nm, power = 200 mW
2. Pin hole diameter	25 μm
3. Spherical lens S	Focal length = 15 cm, optical beam diameter = 22 mm
4. Cylindrical lens (C_1)	Focal length = 30 cm
5. Acousto-optics device (AOD)	Flint glass device, center frequency = 70 MHz, bandwidth = 40 MHz, time aperture = 10 μsec, acoustic beam height = 4mm, distance between AOD2 and AOD3 = 0.2 mm or less
6. Spherical lenses (S_1 and S_2)	Focal length = 50 cm, forms 1:1 imaging system between AOD1 and AOD4 planes, S2 forms an image system along y-direction with a magnification M = 30/50 = 0.6
7. Cylindrical lens (C_2)	Focal length = 20 cm
8. Cylindrical lens (C_3)	Focal length = 30 cm
9. Cylindrical lens (C_4)	Focal length = 10 cm
10. Cylindrical lenses (C_2 and C_4)	Imaging system along x-direction with a magnification M = 10/20 = 0.5

A Mach–Zehnder interferometer method was used for the two-dimensional scanning that uses two crossed Bragg cells as shown in Figure 4.17. A laser light is collimated by a spherical lens and then is split in half by a beam splitter (BS_1). Bragg matched light incident on a Bragg cell AOD1 that produces a diffracted beam that is deflected along the x-direction. A reflecting mirror M_1 is adjusted to cancel the deflection along the x-direction due to the carrier signal such that the diffracted beam is incident on the beam splitter/combiner BS_2. The half signal passes through mirror M_2, which is adjusted to cancel the vertical deflection along the y-direction of the -1 diffracted order from the Bragg cell AOD2 such that it is incident (normal) on the beam splitter/combiner BS_2 and the two beams appear collinear at the output ports of BS_2.

In 1979, Dolfi et al. [22] described a technique in which they focused a single-frequency laser beam through a Bragg cell and the Bragg cell is excited by a continuous

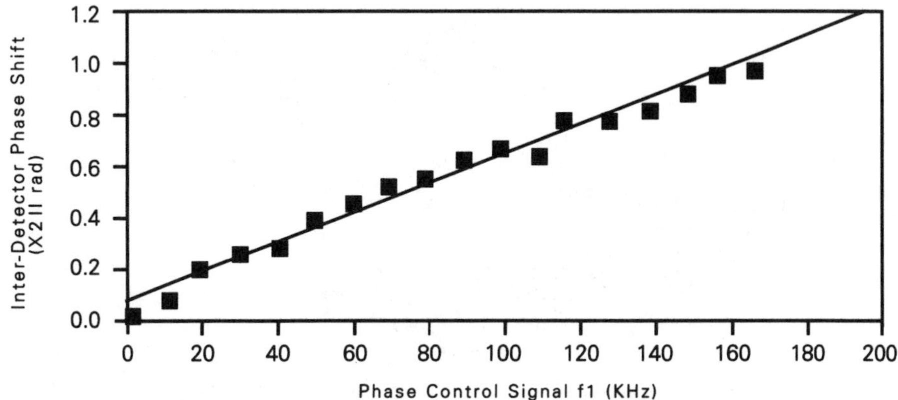

Figure 4.15 Measured phase shift behavior with changing control frequency f_1. (© 1992 SPIE.)

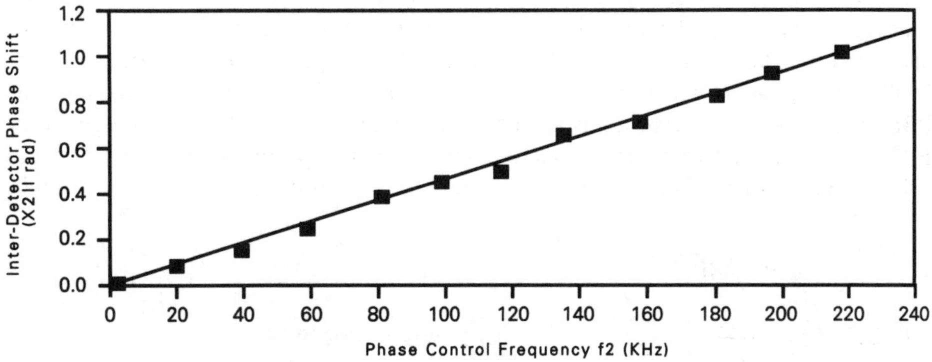

Figure 4.16 Measured phase shift behavior with changing control frequency f_2. (© 1992 SPIE.)

RF/microwave signal and the array antenna beam is steered in one dimension. In 1992, Dolfi et al. [35] described two-dimensional optical architecture for phase and time-delay beamforming for a phased array antenna. In this case, the transmitted beam and diffracted beams are cross-polarized and are recombined without loss on a polarizing beam splitter/combiner (PBS) in order to get a dual-frequency optical carrier of RF/microwave signal. The dual-frequency beam passes through a half-wave plate where the light polarization is rotated by 45° and the beam passes through a beam expander. The expanded beam intercepts M_0, which is a mnematic liquid crystal *spatial light modulator* (SLM) of a $p \times p$ pixel. The polarization of the beam at frequency f_0 concides with the orientation of the liquid crystal molecules. The dual-frequency beam intercepts a set of SLMs, PBSs, and prisms Ps at the output of M_0 as shown in Figure 4.18. Each SLM is made of an array of $p \times p$ pixels, and the beam polarization is rotated by 0° or 90° according to the applied voltage on each pixel. Figures 4.19 and 4.20 show the operating principles of the delays.

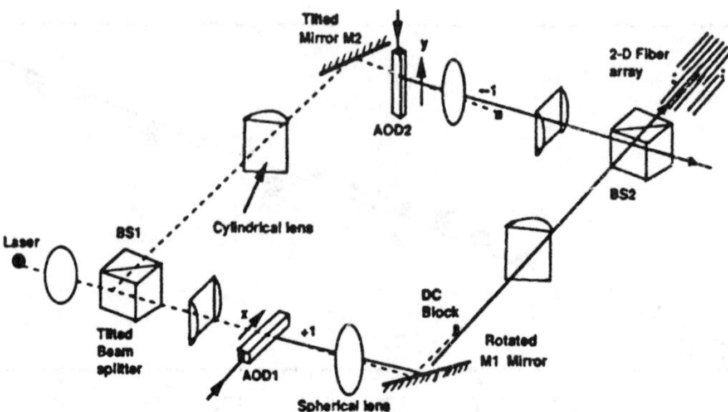

Figure 4.17 Mach–Zehnder architecture for two-dimensional scanning. (© 1991 SPIE.)

The PBS_i is transparent for the horizontal polarized beam and becomes a reflecting device for the vertical polarized beam. The horizontal beam passes through the transparent PBS_i and intercepts the next SLM_{i+1}. The reflected vertical polarized beam from PBS_i is incident on a prism P_i where the beam reflects two sides of P_i and is incident on the second PBS_i, which reflects toward the next SLM. The collimated beam travels through all the PBS_i and is focused by an array of microlenses L onto a $p \times p$ *fiber pigtailed photodiodes* (PDA). The position of the prism P_i provides the delays in beating signal, which can be

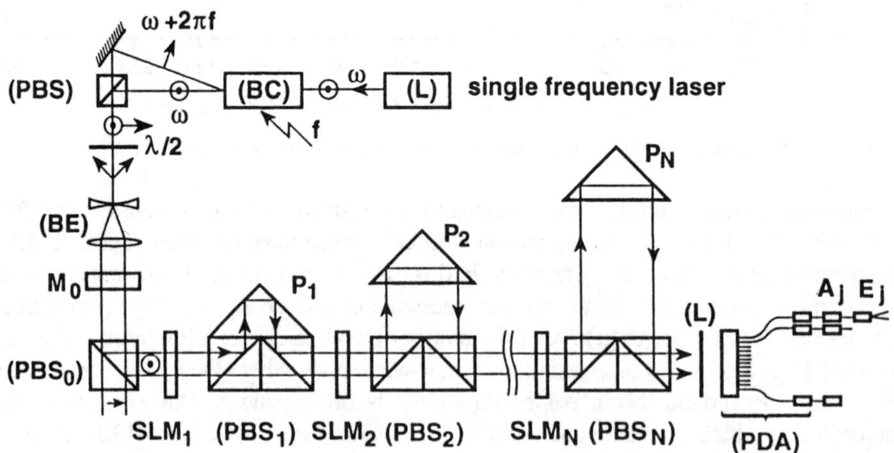

Figure 4.18 Time and phase delays for phased array antenna: (BC) = frequency shifter Bragg cell; (PBS_i) = polarizing beam splitters; (M_0) = spatial light modulator (liquid crystal cell used inthe birefringent mode); (SLM_i) = electrically addressed spatial light modulators; (P_i) = prisms; (L) = lens array; (PDA) = fiber pigtailed photodiode array; (A_j) = microwave amplifiers; and (E_j) = radiating elements.

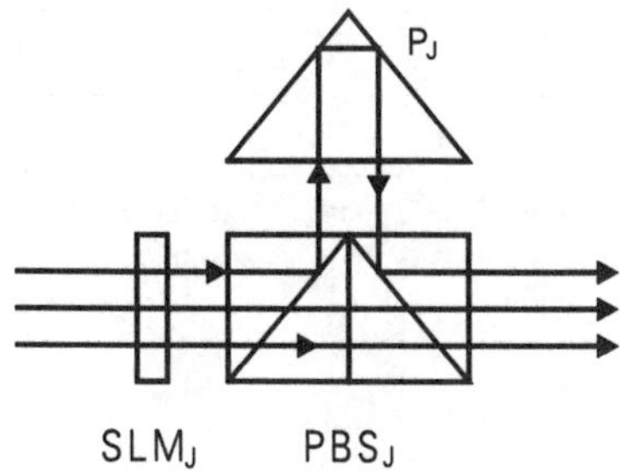

Figure 4.19 Operating principle of delay modules with the position p_j. (© 1992 SPIE.)

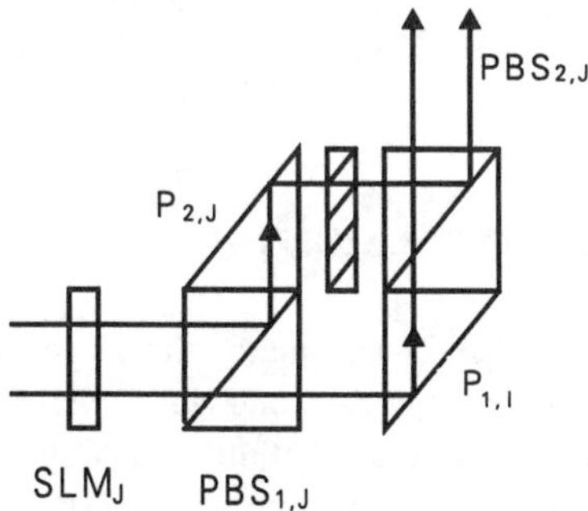

Figure 4.20 Operating principle of delay modules with the thickness of glass. (© 1992 SPIE.)

delayed from 0 to (2^{N-1}) τ with a step equal to τ. The amplitude of the beating signal from each photodetector diode is given by the expression

$$i_k(t) = i_0 \cos\left(2\pi ft + 2\pi f \sum_{j=1}^{N} \varepsilon_{k,j} 2^{j-1}\tau + \frac{2\pi}{\lambda} \cdot e \cdot \Delta n(V_k)\right) \tag{4.29}$$

where λ is the laser wavelength, $i_0 = \sqrt{i_\omega i_{\omega+2\pi f}}$, i_ω(resp. $i_{\omega+2\pi f}$) is the photocurrent delivered by a photodiode detecting the laser beam at frequency $\omega/2\pi$ (resp. $\omega/2\pi + f$) alone, $\Delta n(V_k) = n(V_k) - n_0$, V_k is the applied voltage at each pixel, n is the refractive index, and e is the liquid crystal thickness. The polarizations at the output of SLM_j, $\varepsilon_{k,j} = 1$, when PBS_j reflects the portion of the collimated beam detected by the photodiode k $(1 \le k \le p^2)$ and $\varepsilon_{k,j} = 0$ when PBS_j is transparent. This technique provides $p \times p$ independent channels.

Table 4.2
Summary of the Results for Scanning Antenna Array

Mode Used	Beam Squint	Sidelobe
Phase delays	Beam squint at about 7° scan	No increase
Time delays	No beam squint	No increase
Phase and time delays	No beam squint	Increase in sidelobe due to subarray grating lobes

2^N delay values, and a continuous phase control of the RF/microwave signal between 0 to 2π provided that $e \cdot (n_e - n_0) \geq \lambda$.

Dolfi et al. [35] operated a linear 32-element array to scan in three different modes (pure phase delays, pure time delays, and combine time and phase delay). Their findings are summarized in Table 4.2.

4.8 DISCUSSIONS AND CONCLUSIONS

The method described in this chapter eliminates the large number of electronic phase shifters and their multiple control signals along with extensive hardware and software systems. The Bragg cell that is used to control the optical phase mechanism is dependent on the antenna carrier frequency. The frequency scanning is also used to control the beam of the antenna. The basic in-line design can be made smaller using in-plane counterpropagating acousto-optics devices as shown in Figure 4.21 and described by Lin et al. [29] and Riza [33]. Table 4.3 summarizes the optical efficiency obtained for various acousto-optics techniques.

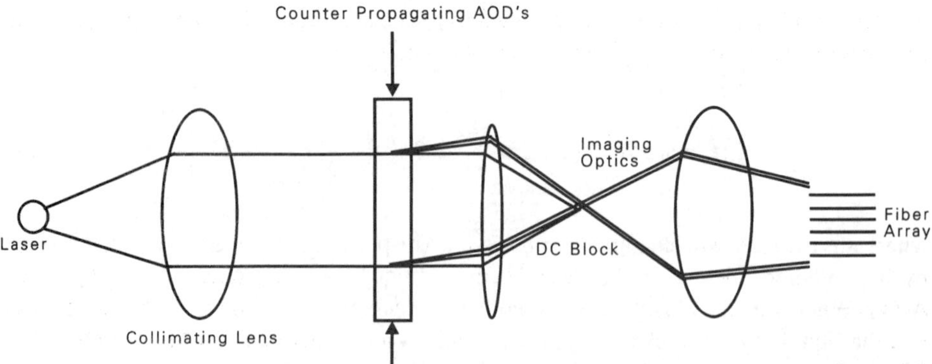

Figure 4.21 Top view of compact in-line architecture for a linear array.

Table 4.3
Optical Efficiency Obtained by Various Acousto-Optic Techniques

Architecture	Optical Efficiency
One-Dimensional Steering	20% to 25%
AOD One-Dimensional Steering	10% to 15%
AOD Two-Dimensional Steering	2% to 5%
Mach–Zehnder Two-Dimensional Steering	5% to 7%

In this chapter, phase-based acousto-optics techniques for controlling the beam in one and two dimensions for linear and planar array antennas have been discussed. The technology is suitable for medium- and larger sized phased arrays for airborne or space-based applications.

References

[1] Brillouin, L., ''Diffusion de la Lumière et des Rayons X par un Corps Transparent Homogene,'' *Ann. Phys. (Paris),* Vol. 17, 1922, pp. 88–122.

[2] Lucas, R., and P. Biquard, ''Propriétés Optiques des Milieux Solides et Liquides Soumis aux Vibration Elatiques Ultra Sonores,'' *J. Phys. Rad.,* Vol. 3, 1932, pp. 464–477.

[3] Debye, P., and F. W. Sears, ''On the Scattering of Light by Supersonic Waves,'' *Proc. Nat. Acad. Sci.* U.S.A., Vol. 18, 1932, pp. 409–414.

[4] Raman, C. V., and N. S. Nagendra Nath, ''The Diffraction of Light by High Frequency Sound Waves,'' *Proc. Ind. Acad. Sci.,* Vol. 2, 1935, pp. 406–420; Vol. 3, 1936, pp. 75–84, 119–125, 459–465.

[5] Nath, N. S. Nagendra, ''Modulation of Light Using Raman-Nath Mode,'' Invited Lecture, Bihar College of Engineering, Patna, Bihar, India, 1965.

[6] Kumar, A., Discussion on Bragg mode and Raman-Nath mode with Dr. N. S. Nagendra Nath, Patna University, Patna, Bihar, India, 1965.

[7] Quate, C. F., C. D. Wilkinson, and D. K. Winslow, ''Intraction of Light and Microwave Sound,'' *Proc. IEEE,* Vol. 53, No. 10, 1965, pp. 1604–1622.

[8] Gordon, E. I., ''A Review of Acousto-Optical Deflection and Modulation Devices,'' *Proc. IEEE,* Vol. 54, No. 10, 1966, pp. 1391–1401.

[9] Adler, R., ''Interaction Between Light and Sound,'' *IEEE Spectrum,* Vol. 4, 1967, pp. 42–53.

[10] King, M., W. R. Bennett, L. B. Lambert, and M. Arm, ''Real Time Electro-Optical Signal Processors with Coherent Detection,'' *Appl. Opt.,* Vol. 6, August 1976, pp. 1367–1375.

[11] Whiteman, R., A. Korpel, and S. Lotsoff, ''Application of Acoustic Bragg Diffraction to Optical Processing Techniques,'' *Proc. Symp. Modern Optics,* 1967, pp. 243–255.

[12] Korpel, A., ''Acousto-optics,'' *Applied Solid State Science,* Vol. 3 (R. Wolfe, ed.), New York: Academic Press, 1972, Ch. 2, pp. 73–179.

[13] Korpel, A., ''Acoustic-Optic Signal Processing,'' *Optical Information Processing* (Yu E. Nesterikhin and G. W. Stoke, eds.), New York: Plenum, 1976, Ch. 10, pp. 171–194.

[14] Korpel, A., ''Acuosto-optics,'' *Applied Optics and Optical Engineering,* Vol. VI (R. Kingslake and B. J. Thompson, eds.), New York: Academic Press, 1980, Ch. 4, pp. 89–136.
[15] Koprel, A., ''Acousto-Optics—A Review of Fundamentals,'' *Proc. IEEE,* Vol. 69, No. 1, 1981, pp. 48–53.
[16] Okoshi, T., ''Hetrodyne and Coherent Optical Fiber Communications: Recent Progress,'' *IEEE Trans. Microwave Theory Tech.,* Vol. MTT-30, No. 8, August 1982, pp. 1138–1149.
[17] Saito, S., Y. Yamamoto, and T. Kimura, ''Optical FSK Hetrodyne Detection Experiments Using Semiconductor Laser Transmitter and Local Oscillator,'' *IEEE J. Qantum. Electron.,* Vol. QE-17, No. 6, June 1981, pp. 935–941.
[18] Bachus, E. J., R. P. Braun, F. Bohnke, G. Elze, W. Eutin, H. Foisel, K. Heimes, and B. Strebel, ''Digital Transmission of TV Signals with a Fiber-Optic Hetrodyne Transmission System,'' *IEEE J. Lightwave Tech.,* Vol. LT-2, No. 4, August 1984, pp. 381–384.
[19] Kumar, A., ''Acousto-Optics Delay Line Concept,'' Research EETC, Coventry, U.K., 1982.
[20] Koepf, G. A., ''Optical Processor for Phased-Array Antenna Beam Formation,'' *Proc. SPIE,* Vol. 477, 1984, pp. 75–81.
[21] Chang, I. C., and S. S. Tarng, ''Phased Array Beamforming Using Acousto-Optic Techniques,'' *Proc. SPIE,* Vol. 936, 1988, pp. 163–167.
[22] Dolfi, D., J. P. Huignard, and M. Baril, ''Optically Controlled True Time Delays for Phased Array Antenna,'' *Proc. SPIE Optical Tech. Microwave Appl. IV,* Vol. 1102, 1989, pp. 152–161.
[23] Toughlin, E. N., H. Zmuda, and P. Kornreich, ''A Deformable Mirror-Based Optical Beamforming System for Phased Array Antennas,'' *IEEE Photonics Tech. Lett.,* Vol. 2, June 1990, pp. 444–446.
[24] Toughlin, E. N., and H. Zmuda, ''Variable RF Delay Line for Pased Array Antennas,'' *IEEE J. Lightwave Tech.,* Vol. 8, No. 12, December 1990, pp. 1824–828.
[25] Dolfi, D., F. Michel-Gabriel, S. Bann, and J. P. Huignard, ''Two-Dimensional Optical Artitecture for Time Delay Beamforming in a Phased Array Antenna,'' *Optics Lett.,* Vol. 16, February 1991, pp. 255–257.
[26] Gesell, L. H., and J. L. Lafuse, True Time Delay Beam Formation, U.S. Patent Application Serial No. 07/806,697, December 1991.
[27] Gesell, L. H., and T. M. Turpin, ''True Time Delay Beamforming Using Acousto-Optics,'' *Proc. SPIE,* Vol. 1703, 1992, pp. 592–602.
[28] Zmuda, H., and E. N. Toughlin, ''Adaptive Microwave Signal Processing: A Photonic Solution,'' *Microwave J.,* Vol. 35, No. 2, February 1992, pp. 58–71.
[29] Lin, S. C., and R. S. Boughton, ''Acousto-Optic Multichannel Programmable True Time Delay Lines,'' *Proc. SPIE, Optical Tech. Microwave Appl.* IV, Vol. 1102, 1989, pp. 162–173.
[30] Riza, N. A., ''Acusto-Optic Techniques for Phased Array Antenna Processing,'' *Proc. SPIE Conf. Emerging Optoelec. Tech.,* Vol. 1622, No. 21, 1991.
[31] Riza, N. A., ''Acousto-Optic Architectures for Multi-Dimensional Phased Array Antenna Processing,'' *Proc. SPIE, Optical Tech. for Microwave Appl.* V, Vol. 1476, 1991, pp. 144–156.
[32] Riza, N. A., ''An Acousto-Optic Phased Array Antenna Beamformer with Independent Phase and Carrier Control using Single Sideband Signals,'' *IEEE Photonics Tech. Lett.,* Vol. 4, No. 2, February 1992.
[33] Riza, N. A., ''High Speed Two Dimensional Phased Array Antenna Scanning Using Acousto-Optics,'' *Proc. SPIE,* Vol. 1703, 1992, pp. 460–468.
[34] Aulock, W. H., ''Properties of Phased Arrays,'' *IRE Trans. Ant. Prop.,* Vol. AP-9, 1960, pp. 1715–1727.
[35] Dolfi, D., S. Bann, J. P. Huignard, and J. Roger, ''Two-Dimensional Optical Architecture for Phase and Time-Delay Beam Forming in a Phased Array Antenna,'' *Proc. SPIE,* Vol. 1703, 1992, pp. 481–489.

Chapter 5

Optically Controlled Beam Scanning of Active Phased Array Antennas

5.1 INTRODUCTION

The active integrated antenna is a combination of active devices and passive antenna elements on the substrate. Semiconductor devices and printed circuit antennas are usually used as the active devices and passive antenna elements, respectively. In recent years, the application of the active integrated antenna in the beam-steering technique has become an area of growing interest due to mature technology of *microwave integrated circuit* (MIC), *monolithic microwave integrated circuit* (MMIC), and the active sources (two terminal devices, such as IMPATT diode and Gunn diodes, as well as three terminal devices, such as MESFETs, HEMTs, and HBTs). Generally, microstrip resonant antennas such as patch and slot antennas are commonly used as the radiating elements. They are not only the output loads of oscillators but also serve as the resonators that determine oscillation frequencies. The input impedance of the antenna element provides important information for the design of oscillator-type active integrated antennas. By using the modern MIC and MMIC fabrication technology, compact, light-weight, and low-cost active integrated beamforming, active antenna systems are realized. An active integrated antenna array can function as a phased array. The phase of the array is controlled by the following techniques: (1) application of phase shifters, and (2) methods of injection locking.

5.2 APPLICATION OF PHASE SHIFTERS

The principles of beam scanning in phased arrays are reported in many books [1–5]. In a conventional phased array, a constant phase progression is established using electronically controlled phase shifters at each array element as discussed in Chapter 3 at RF/microwave frequencies.

An optoelectronic phased array antenna architecture using an externally modulated laser and phase shifter technique for steering the antenna beam is described. The active phased antenna array system consists of radiating elements, amplifiers, detectors, and phase shifters. Figure 5.1 shows a block diagram of the phased array antenna architecture. A laser diode is modulated to form a microwave envelope on the optical carrier using an external modulator. Data is impressed on the optical signal by using a switched optical delay line that is made up of discrete delays, allowing phase modulation of the carrier envelope. This optical signal is split by a star divider, delayed by a multistate optical delay line to achieve microwave beam steering, and distributed via an optical line to each transmit element. This discrete optical delay may be fabricated from waveguides of in-diffused lithium niobate or discrete lengths of fiber. The phase/data modulation and the phase delays to achieve beamforming and steering are accomplished by similar if not identical methods. The transmit model at the antenna consists of an optical detector that recovers the phase-shifted, *phase-shifted-keyed* (PSK) signal. The beam control channels provide appropriate signal information to phase shifters to steer the beam.

A *piezoelectric* (PZT) crystal-based phase shifter provides a better alternative to steer the beam of active arrays than conventional phase shifters. In this case, fibers are wrapped in groups of four or eight around PZT crystals. The capacitive breakdown voltage of the PZT crystal is calculated to be as high as 4 kV. The biasing voltage provides the expansion of the PZT ring. The expansion of the PZT ring causes the fiber to stretch, which introduces a variable time delay for the phase-reference signal. The phase shift ($\Delta\Phi$ in radians) due to a stretch in the fiber is given by [6, 7]

$$\Delta\Phi = 2\pi(\Delta l/\lambda_g) \tag{5.1}$$

where Δl is the stretch in fiber and λ_g is the guide wavelength of the microwave modulating envelope of the optical carrier.

Equation (5.1) can be written in terms of PZT parameters as

$$\Delta\Phi = 2N\pi^2 D\ 10^{-5}\ V/\lambda_g \tag{5.2}$$

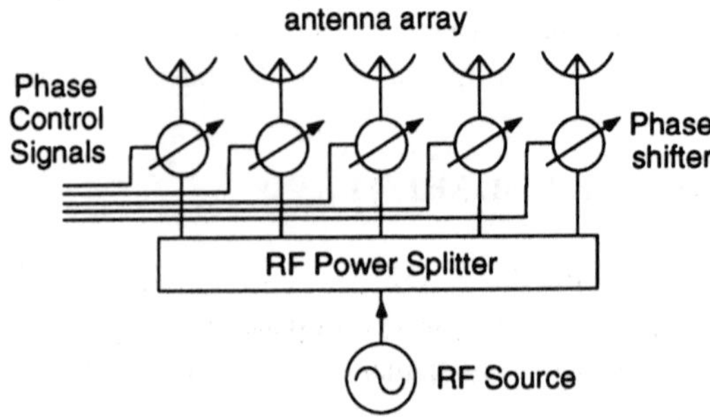

Figure 5.1 Block diagram of the phased array antenna architecture using phase shifters.

where N is the number of turns of fiber, D is the diameter of the PZT ring, and V is the DC bias voltage in kilovolts. Daryoush et al. [6] investigated the PZT crystal time-delay phase shifter by analyzing on an HP8510 network analyzer. The phase of the transmission S-parameter (S_{21}) was measured for the case of 0 DC biasing, where a reference phase was established. Then, upon application of a DC biasing voltage, the phase of S_{21} was measured and compared to the reference phase. Table 5.1 shows the transmission phase shift as a function of a DC biasing voltage. It is anticipated that the optical fiber may be stretched repeatedly to 5% of its length.

Generally, the beam characteristics of a uniform linear array are controlled by the amplitudes (gains) and phase shifts of the signals routed to the radiating elements, which are shown in Figure 5.2. Kam et al. [7] reported a small number of shared phase shifters

Table 5.1
Transmission Phase Shift ΔS_{21} as a Function of DC Biasing

DC biasing voltage (kV)	0	1.6	2.4	3.2	4.2
ΔS_{21} phase shift (degrees)	Reference	−7	−12	−16	−20

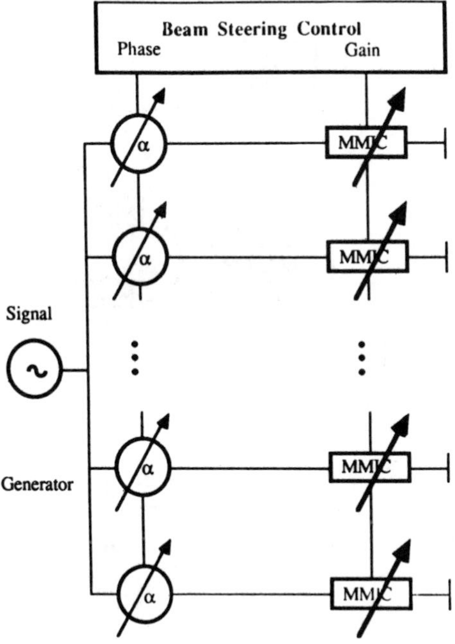

Figure 5.2 A steerable linear array with phase shifters. (© 1989 IEEE.)

in the signal distribution and phasing subsystem in order to steer the beam, as shown in Figure 5.3. Figure 5.4 shows a block diagram of an optically controlled phase array antenna system. In the figure, the microwave signals from the primary signal generator are converted to the optical domain and pass through the optical processor (PZT crystal phase shifter) to the front-end MMICs and arrays. Figure 5.5 shows a PZT crystal-based phase shifter for a four-element antenna array. The DC biasing voltage controls the phase shift of the antenna elements.

Two architectures for a 16-element array are described that are based on a uniform linear array with a separate phase shifter for each element (Figure 5.6(a)) and shared phase shifter for a subgroup of four elements (Figure 5.6(b)). All phase shifters in Figure 5.6(a) are statistically uncorrelated due to the independence of the paths. In Figure 5.6(b), fully correlated phase errors are provided by the optical paths that share the same phase shifter (for example, the second and the third). However, paths that use different phase shifters (for example, the second and the fifth) are statistically uncorrelated. Kam et al. [7] conducted an analysis on the effects of phasing errors on beam direction. It is demonstrated that beam-steering errors in the shared phase shifter array architectures can be reduced significantly by using an appropriate sharing scheme.

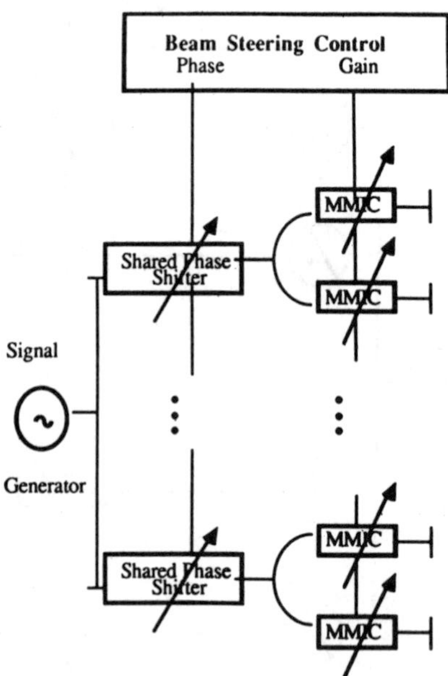

Figure 5.3 A steerable linear array with shared phase shifters. (© 1989 IEEE.)

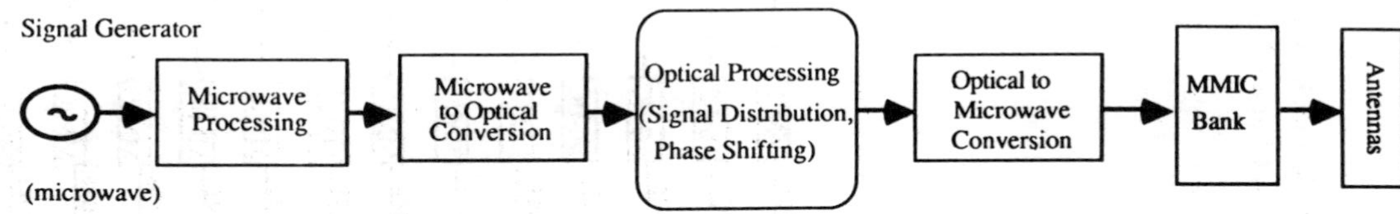

Figure 5.4 An optically controlled phased array antenna system. (© 1989 IEEE.)

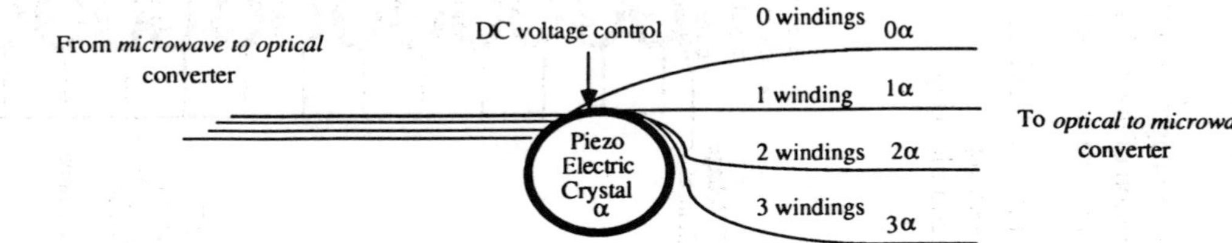

Figure 5.5 PZT crystal-based phase shifter for four elements. (© 1989 IEEE.)

5.3 METHODS OF INJECTION LOCKING

In oscillator-type active integrated antenna arrays, injection locking is an important technique. A short description of this technique is available in Chapter 3. There are three types of injection locking that we use in this chapter:

1. Subharmonic injection locking;
2. Interinjection locking;
3. Unilateral injection locking.

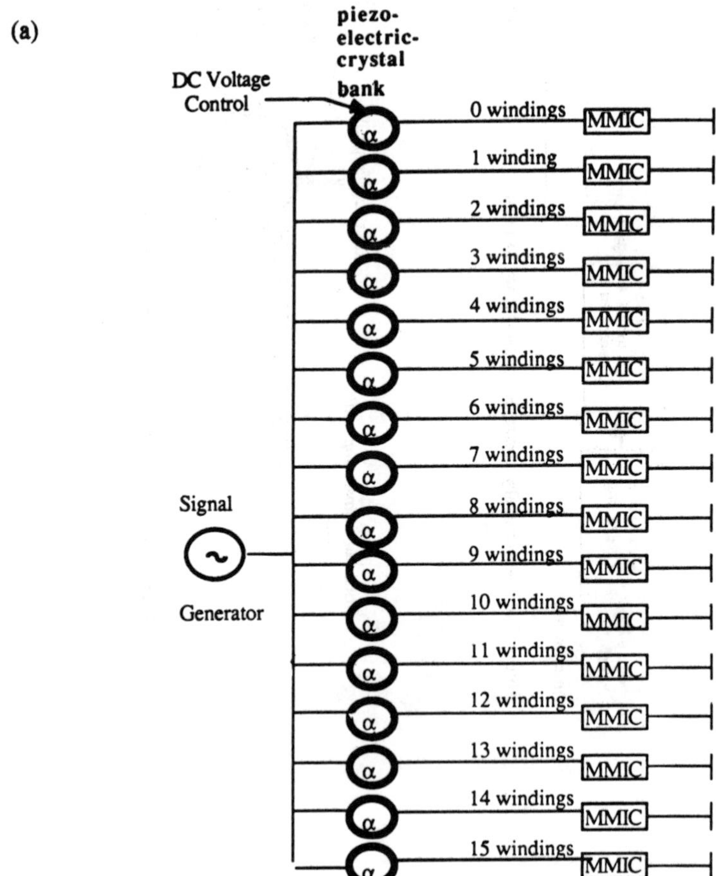

Figure 5.6 A 16-element optically controlled array with (a) 16 independent phase shifters and (b) 4 shared phase shifters. (© 1989 IEEE.)

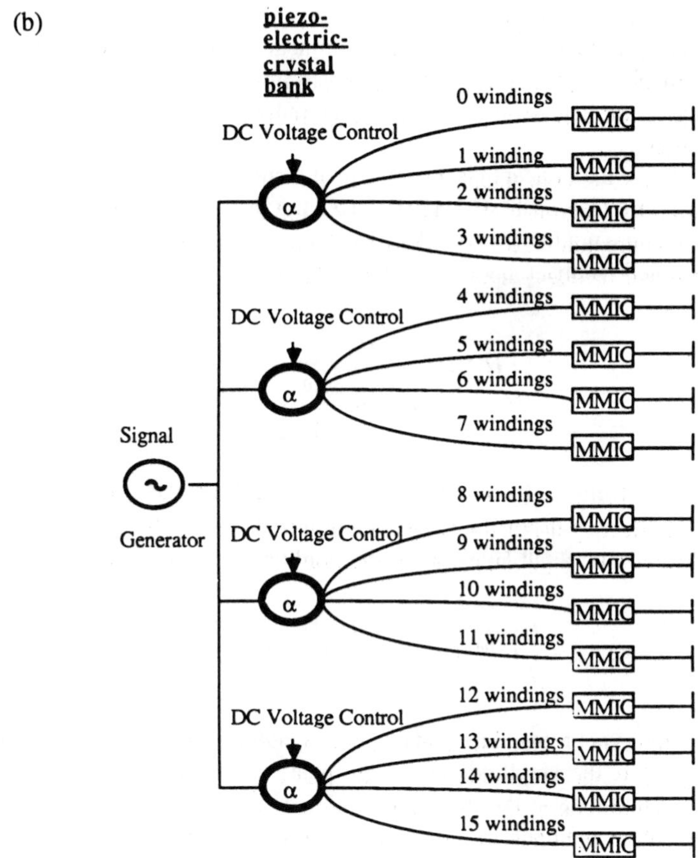

Figure 5.6 (continued)

5.3.1 Subharmonic Injection Locking

Instead of using phase shifters, a method of using subharmonic injection locking to establish a progressive phase is proposed by Daryoush [8] and Daryoush et al. [9]. They demonstrated the application of subharmonic injection locking for optical synchronization of remotely located oscillators at millimeter-wave frequencies. The nonlinear characteristics of solid-state devices and the effect of oscillator circuit technology on the subharmonic injection locking figures of merit are critical in establishing an efficient synchronization. A general nonlinear model for injection locking oscillators that is based on Van der Pol's equivalent circuit model representation is discussed by Adler [10]. This type of circuit provides one-port topology. Adler assumed the small perturbation signal in the solution; therefore, it cannot easily be implemented for microwave oscillators. In 1991,

Berceli et al. [11] reported a two-port model to analyze the subharmonic injection-locking range of a microwave oscillator in terms of the nonlinear current–voltage relationship of an active device. Zhang et al. [12] provided the analysis of a subharmonic injection-locked local oscillator based on a general nonlinear input–output model for the subharmonic synchronized oscillator.

Figure 5.7 shows the conceptual diagram of the subharmonic synchronized oscillator that is reported by Daryoush et al. [13] and Zhang et al. [14]. The model provides the combination of a pure nonlinear network *f(e)* and a pure linear feedback network *H(D)*. The linear single-tuned feedback network *H(D)* is reported by Schmideg [15] as

$$H(D) = \frac{H_0}{1 + j2Q\dfrac{\Delta\omega}{\omega_0}} \tag{5.3}$$

where Q is the quality factor of the nonlinear network and $\Delta\omega$ is the frequency deviation from the resonant frequency ω_0 of the feedback circuit. When a signal e_i is injected and the oscillator is locked, the input signal e for the nonlinear network is given by

$$e = e_0 + e_i = \frac{E}{2}\left(e^{j\omega t} + e^{-j\omega t}\right) + \frac{\dot{E}_i}{2}e^{j(\omega/n)t} + \frac{\dot{E}_i^*}{2}e^{-j(\omega/n)t} \tag{5.4}$$

where $\omega = n\omega_{\text{inj}}$ is the synchronized frequency after injection locking, ω_{inj} is the injection angular frequency, E_i is the amplitude and ϕ the phase of the injected signal, E is the oscillation signal's amplitude at the input port, and n is an integer for the subharmonic factor. The input–output nonlinearity of the active device is represented by *f(e)* and is expressed as

$$u = f(e) = \sum_{m=-\infty}^{\infty} \dot{U}_m e^{jm(\omega/n)t} \tag{5.5}$$

where α_i is assumed to be real for simplicity. The output of the oscillator is given by

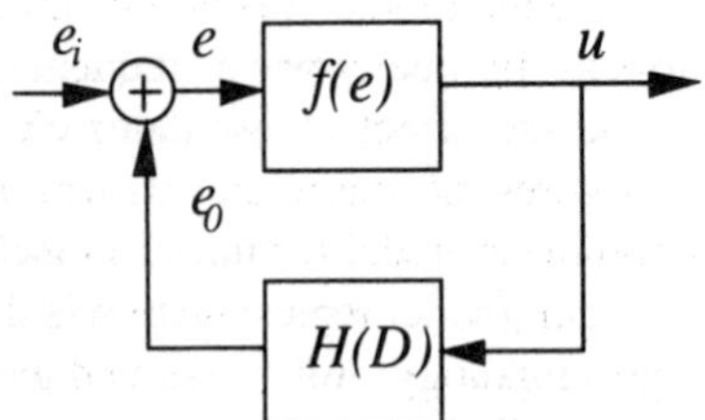

Figure 5.7 Block diagram of subharmonically synchronized oscillator. (© 1992 IEEE.)

$$u = f(e) = \sum_{i=1}^{\infty} \alpha_i e^i \tag{5.6}$$

An output signal U_n at the oscillation frequency $n\omega_{inj}$ is calculated by substituting (5.6) into (5.5), yielding

$$\dot{U}_n = \left(\sum_{j=0}^{\infty} \sum_{k=0}^{\infty} \frac{1}{2^{N-1}} \frac{N!}{(j!)^2(k+1)!k!} \alpha_N |E_i|^{2j} E^{2k} \right) E$$
$$+ \left(\sum_{m=0}^{\infty} \sum_{p=0}^{\infty} \frac{1}{2^{M-1}} \frac{M!}{(m!)^2(p+n)!p!} \alpha_M |E_i|^{2p} E^{2m} \right)$$
$$\cdot \dot{E}_i^n + \text{higher order terms} \tag{5.7}$$

where $N = 2j + 2k + 1$ and $M = 2m + 2p + n$.

Equation (5.7) can be rewritten as

$$\dot{U}_n \approx U_{out} + \dot{U}_{outn}. \tag{5.8}$$

where

$$U_{out} = \left(\sum_{j=0}^{\infty} \sum_{k=0}^{\infty} \frac{1}{2^{N-1}} \frac{N!}{(j!)^2(k+1)!k!} \alpha_N |E_i|^{2j} E^{2k} \right) E$$

$$\dot{U}_{outn} = \left(\sum_{m=0}^{\infty} \sum_{p=0}^{\infty} \frac{1}{2^{M-1}} \frac{M!}{(m!)^2(p+n)!p!} \alpha_M |E_i|^{2P} E^{2m} \right) \dot{E}_i^n$$

The oscillation signal E can be simplified as

$$E = H_0 U_{out} \tag{5.9}$$

and $\Delta\omega$ is given by

$$\Delta\omega = \frac{\omega_0}{2Q} \frac{\text{Im}\,(\dot{U}_n)}{\text{Re}\,(\dot{U}_n)} \approx \frac{\omega_0}{2Q} \frac{\sin\,(n\phi)\,|\dot{U}_{outn}|}{U_{out}} \tag{5.10}$$

The subharmonic injection-locking range can be expressed in terms of Q and ω_0 by the method of Daryoush et al. [9] and assuming $n\phi = \pm\pi/2$ as

$$\Delta\omega_{1/n} \approx \frac{\omega_0}{2Q} \frac{U_{outn}}{U_{out}} = \frac{\omega_0}{2Q} \sqrt{\frac{P_{outn}}{P_{out}}} \tag{5.11}$$

The above equation is the same as Adler's [10] equation for a fundamental injection-locking range when the signal interacts with the free-running oscillator like a fundamental locking signal.

In the present system, the injection signal passes through the nonlinear network and its nth harmonic signal is generated at the output part of the network. When the injection frequency at $n\omega_{inj} \approx \omega_0$, the multiplied signal can be fed back to the nonlinear network again through the linear tank circuit. This signal makes the free-running oscillator synchronized at $\omega = n\omega_{inj}$, the same as that for a fundamental locking signal.

To verify the analysis of injection locking, a two-stage low-noise MESFET amplifier and a dielectric resonator are constructed into a linear feedback network. The measured and calculated injection-locking ranges ($\Delta\omega/\omega_0$) at subharmonic factors $n = 2$, 3, and 4 are shown in Figure 5.8. In the figure, the calculation for Model 1 represents the calculated values from (5.11), and the calculation for Model 2 represents the predicted results with unlocked spectrum as suggested by Bussgang et al. [16].

The injection-locking phenomenon depends upon the spectral noise density of the pilot carrier within the passband of the oscillator, giving rise to an average frequency due to the FM noise jitter of the injection-locked, free-running oscillator. The close-in carrier phase noise of the subharmonically injection-locked oscillator has been determined by considering the case where the FM noise is close to the carrier, that is,

$$\Omega << \Delta\omega_{1/n}$$

In the above case, $\Delta\omega_{1/n}$ represents the locking range and Ω represents the offset frequency. The output FM noise/carrier ratio $£_{1/n}(\Omega)$ of an nth-order subharmonically locked local oscillator can be formulated in terms of the injection FM noise/carrier ratio [14, 17] as

$$£_{1/n}(\Omega) \approx n^2 £_{inj}(\Omega) + \frac{(1 + \omega_c/\Omega)}{\Delta\omega_{1/n}^2} \Delta\Omega^2 \tag{5.12}$$

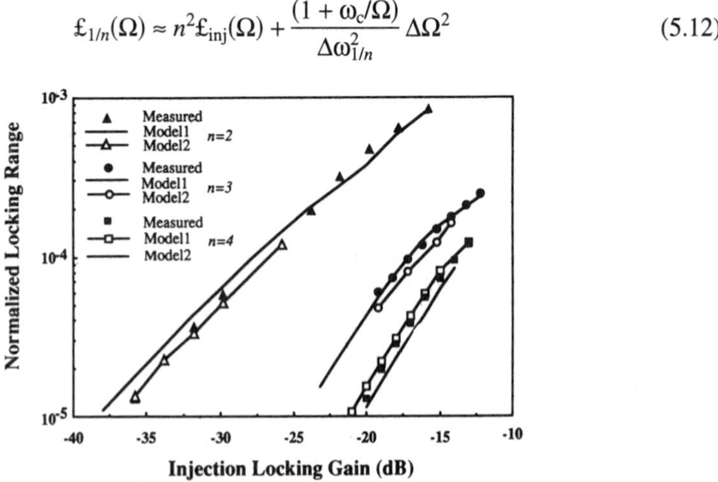

Figure 5.8 Plots of experimental and theoretical results of normalized locking range versus injection-locking gain at subharmonic factors n = 2, 3, 4, where theoretical values for model 1 and model 2 are calculated from [14] and [16], respectively. (© 1992, 1974 IEEE.)

where $£_{inj}(\Omega)$ is the spectral noise power density of the pilot carrier, ω_c is the corner frequency for a $1/f$ near-carrier noise of the device, and $\Delta\Omega$ represents the frequency jitter and is calculated from the expression

$$\Delta\Omega = \frac{\omega_0}{2Q}\sqrt{\frac{P_{outn}}{P_{out}}} \tag{5.13}$$

Equation (5.13) is similar to (5.11) for the injection-locking range calculation, where P_{outn} is the injection power and P_{out} is the output power. So, it can be considered as an average of the frequency shift by the average noise power within 1 Hz through the locking process.

The second term in (5.12) provides intrinsic noise when $\Delta\omega_{1/n}$ is small. In this case, the relation of the output FM noise with the locking range is approximately $1/x^2$; therefore, when the input power increases, $\Delta\omega_{1/n}$ increases and the output FM noise decreases. Zhang et al. [14] showed this phenomenon in Figure 5.9, which provides measured and calculated phase noise at 1-kHz offset carrier frequency of the injection-locked oscillator at subharmonic factors of 1/2, 1/3, and 1/4, as a function of the injection-locking range. The figure shows that the second term becomes very small when $\Delta\omega_{1/n}$ is large, which makes $£_{1/n}(\Omega)$ approach a minimum value of $n^2£_{inj}(\Omega)$.

The minimum FM noise degradation factor is n^2 for an nth-order subharmonic injection-locking oscillator. The intersection points A, B, and C between the line of $1/x^2$ and the limit lines represent 3-dB turning points. These points show the locking range or injection power level needed to obtain the FM noise level, which is 3 dB higher than the minimum level. The locking range of a 3-dB turning point is given by

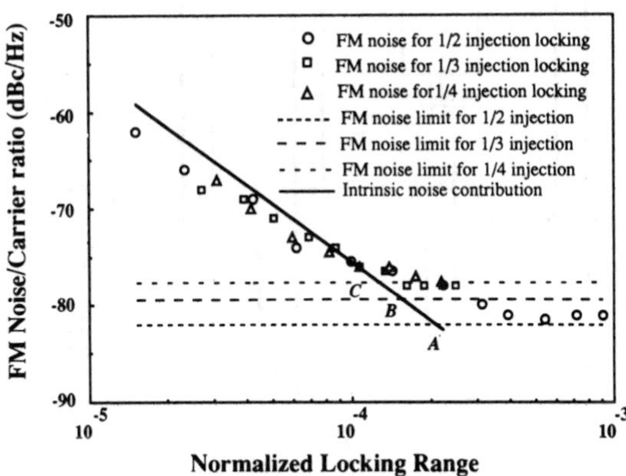

Figure 5.9 Experimental and theoretical values of phase noise at 1-kHz offset carrier frequency of the injection-locked oscillator at subharmonic factors of 1/2, 1/3, and 1/4, as a function of the injection-locking range. (© 1992 IEEE.)

$$\Delta\omega_{1/n} \approx \frac{\sqrt{1 + \omega_c/\Omega}}{\sqrt[n]{\pounds_{\text{inj}}(\Omega)}} \Delta\Omega \tag{5.14}$$

The injection power needed to reach a 3-dB turning point is calculated by the expression

$$P_{\text{inj}} \approx \frac{1}{b^2 n^2}\left(1 + \frac{\omega_c}{\Omega}\right)\frac{P_N}{\pounds_{\text{inj}}(\Omega)M} \tag{5.15}$$

where M is the nonlinear network multiplication factor for the subharmonic injection signal and R_N is the intrinsic noise level.

5.3.2 Beam Steering Using Subharmonic Injection Locking

The MMIC technology has made feasible the use of very large phased array antennas in space applications where once weight and complexity precluded its use. The concept of an optically controlled T/R model phased array antenna has been reported by many researchers [17–23]. A subharmonic optically injection-locked oscillator can provide a precise analog phase shift of −90° to +90° [20–22] with minimal phase noise degradation. This limits the steering of the antenna beam. The phase shift can be increased from −180° to +180° (i.e., a full 360°) by the system described by Zhang et al. [24]. In their system two cascade suboscillators are utilized in the circuit as shown in Figure 5.10. In the circuit,

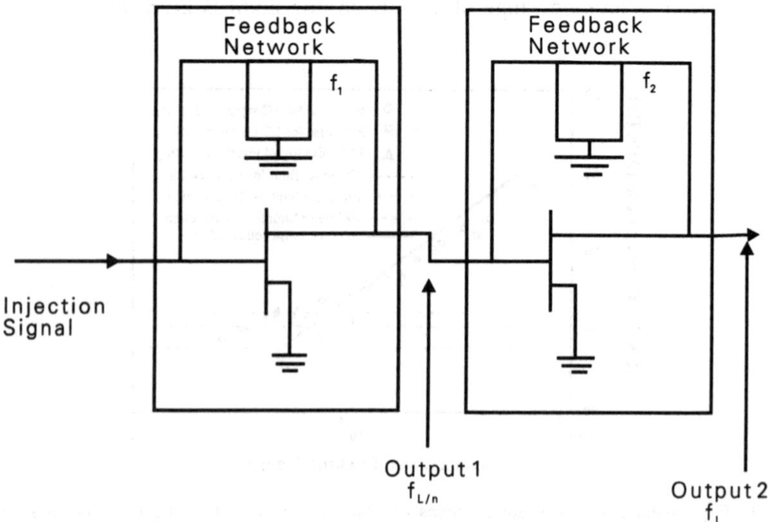

Figure 5.10 Block diagram of the 360° phase shifted injection-locked oscillator. (© 1993 IEEE.)

a tunable tank network is connected to a transistor-based gain stage with a positive feedback to generate oscillation in each suboscillator. The first suboscillator oscillates at the nth subharmonic of the oscillation frequency of the second suboscillator. In the system, f_1 and f_2 are free-running frequencies of the first and the second suboscillators. The reference (injection) signal locks the first suboscillator subharmonically, and then the locked output from the first suboscillator locks the second suboscillator at a subharmonic factor of n to generate a stable local oscillator signal. In the case of injection locking, the $f_{L/n}$ and f_L are locked frequencies for the two cascade oscillators. The detuning frequencies of δf_1 and δf_2 exist between the locked and the free-running suboscillators. The phase shifts at the first suboscillator ($\Delta\phi_1$) and the second suboscillators ($\Delta\phi_2$) are given by

$$\Delta\phi_1 = \arcsin\left(\frac{2\delta f_1}{\Delta f_1}\right) \tag{5.16}$$

$$\Delta\phi_2 = \arcsin\left(\frac{2\delta f_2}{\Delta f_2}\right) \tag{5.17}$$

where Δf_1 and Δf_2 are the subharmonic injection-locking ranges of the two cascade suboscillators. The total phase shift at output of the local oscillator circuit for a fixed injection frequency and phase is written as

$$\begin{aligned}\Delta\phi_{LO} &= n\Delta\phi_1 + \Delta\phi_2 \\ &= n \arcsin\left(\frac{2\delta f_1}{\Delta f_1}\right) + \arcsin\left(\frac{2\delta f_2}{\Delta f_2}\right)\end{aligned} \tag{5.18}$$

The first and second terms of the above equation provide the range of phase variation of $n \times 180°$ and $180°$, respectively. These values are possible if we make Δf_1 and Δf_2 fixed and assume change in δf_1 and δf_2. Therefore, a phase shift of 360° is observed. Zhang et al. [24] verified the concept, and Kumar [25] built a similar system in the laboratory. We measured a phase shift of about 310° over the locking range. To stabilize the free-running frequencies of the two oscillators, dc-phase-lock-loops are used and a phase shift of about 350° is observed instead of 360°. This result is very close to the theoretical value. Recently, this design has been successfully used in the design of a large satellite antenna [25] in place of a switched delay line phase shifter in T/R modules.

5.3.3 Interinjection-Locking Method

Interinjection locking may be defined as the power-combining problem by distributing power sources among the individual antennas of the array. Phase control is effected largely through interactions among oscillators. The dynamics of the injection-locking process establishes a constant phase progression across an array of coupled oscillators when the end elements are detuned in a particular way. The main advantage of this technique is that there is no need of a feed network. Stephan [26] and Stephan et al. [27, 28] proposed a coupled oscillator for injection locking. In their approach, two signals with a controlled phase difference, $\Delta\phi$, are injected into opposite ends of the array. The resulting phase difference between each of the N oscillators is then found to be $\Delta\phi/(N+1)$; therefore, increasing the number of oscillators in the array decreases the maximum available phase difference between each oscillator. In this case, the scanning range is quite limited for even the modest-sized array.

The element phase shift obtained in this interinjection technique [29] is independent of the number of oscillators. This technique utilizes an array of individual solid-state oscillators that are integrated with planar antennas. Mutual coupling between the antenna elements allows the oscillators to interact and to synchronize with a common frequency via injection locking.

A theory of coupled oscillators predicts the steady-state phase relationships in the arrays. In the case of power combining the oscillator array, York et al. [30–33], York [34], and Liao et al. [35, 36] described the array by coupled Vander Pol equations. They presented a single oscillator by an RLC circuit, with a voltage source to represent injected signals, and a negative resistive device. The phase dynamics of N coupled oscillators with nearest neighbor coupling are given by

$$\frac{d\theta_i}{dt} = \omega_i - \frac{\omega_i}{2Q}\sum_{j=1}^{N} \varepsilon_{ij}\frac{A_j}{A_i}\sin\left(\Phi_{ij} + \theta_i - \theta_j\right) \tag{5.19}$$

where $i = 1, 2, 3, \ldots, N$, ω_i is the free-running frequency, θ_i is the instantaneous phase, A is the amplitude of oscillation, Q is the quality factor of the ith oscillator, ϵ and Φ are the magnitude and phase of the coupling, and subscripts i and j denote coupled oscillators i and j. The mutual interaction between oscillators is described by a complex coupling between oscillators i and j and is given by $\epsilon_{ij}e^{j\Phi_{ij}}$ · ϵ_{ij} and Φ_{ij} are the amplitude and the phase of the coupling due to mutual interaction between oscillators i and j.

In the case of weak coupling, $\epsilon_{ij} \ll 1$ and the amplitudes of the oscillators will remain close to their free-running values, that is, $A_i = \alpha_i$ and $A_j = \alpha_j$. The phase dynamics of the system is given by simplifying the equation as follows:

$$\frac{d\theta_i}{dt} = \omega_i - \frac{\omega_i}{2Q}\sum_{j=1}^{N} \varepsilon_{ij}\frac{\alpha_j}{\alpha_i}\sin(\Phi_{ij} + \theta_i - \theta_j) \qquad i = 1,2,\ldots,N \tag{5.20}$$

Figure 5.11 shows a linear array of N active oscillators with the nearest neighbor coupling described by

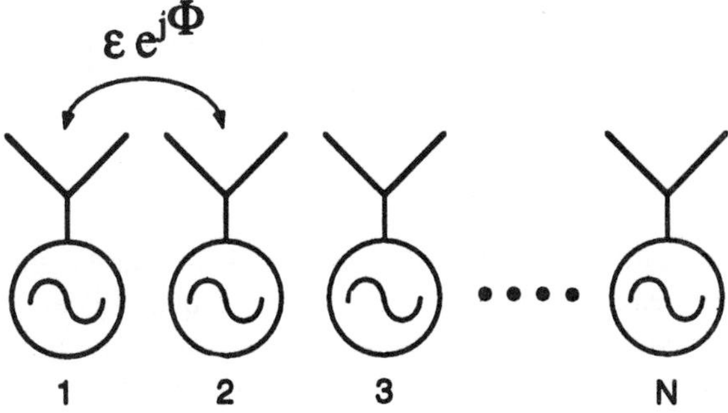

Figure 5.11 Sketch of a linear array of N active antenna oscillators with nearest-neighbor coupling described by a coupling strength and coupling angle. (© 1993 IEEE.)

$$\kappa_{ij} = \begin{cases} \varepsilon e^{-j\Phi} & \text{if } |i - j| = 1 \\ 0 & \text{otherwise} \end{cases} \tag{5.21}$$

where κ_{ij} is a complex coupling coefficient due to mutual interaction between oscillators i and j. From (5.20) and (5.21), we can write

$$\frac{d\theta_i}{dt} = \omega_i - \frac{\varepsilon\omega_i}{2Q} \sum_{\substack{j=i-1 \\ j \neq 1}}^{i+1} \frac{\alpha_j}{\alpha_i} \sin(\Phi + \theta_i - \theta_j)$$
$$i = 1, 2, \ldots, N \tag{5.22}$$

where $\alpha_0 = \alpha_{N+1} \equiv 0$. When the oscillators mutually lock, we can make the assumption that

$$\Delta\omega_m \equiv \frac{\varepsilon\omega_i}{2Q} \tag{5.23}$$

From Adler's equation [10] of injection locking, we can rewrite the locking bandwidth of the oscillator as

$$\Delta\omega_{\text{lock}} = \frac{\omega_0}{2Q} \frac{A_{\text{inj}}}{A} \tag{5.24}$$

By comparing (5.23) and (5.24), the value of $\Delta\omega_m$ in (5.23) can be interpreted as the locking range of the ith oscillator provided all the amplitudes are the same. We can replace the value of free-running frequencies $\epsilon_i\omega_i/2Q$ by $\Delta\omega_m$ in (5.22) because the free-running frequencies should be very close to $\Delta\omega_{\text{lock}}$ for frequency locking to occur. The value of $d\theta_i/dt$ is also replaced by ω_f, and (5.22) can be rewritten as

$$\omega_f = \omega_i \left[1 - \frac{\varepsilon}{2Q} \sum_{\substack{j=i=1 \\ j \neq i}}^{i+1} \frac{A_j}{A_i} \sin(\Phi + \theta_i - \theta_j) \right] \tag{5.25}$$

$$i = 1, 2, \ldots, N$$

The steady-state phase differences between each oscillator can be solved using these N equations. Electronic beam scanning in antenna arrays requires a constant phase progression along the array and $\theta_i - \theta_{i-1} = \Delta\theta$ for all i. A constant phase progression $\Delta\theta$ can be synthesized at a frequency ω_f for the following distribution of free-running frequencies:

$$\omega_i = \begin{cases} \omega_f[1 + \varepsilon' \sin(\Phi + \Delta\theta)] & \text{if } i = 1 \\ \omega_f[1 + 2\varepsilon' \sin\Phi \cos\Delta\theta] & \text{if } 1 < i < N \\ \omega_f[1 + \varepsilon' \sin(\Phi - \Delta\theta)] & \text{if } i = N \end{cases} \tag{5.26}$$

This requires that all of the innermost oscillators share the same free-running frequency, the end elements should be slightly detuned. The progressive phase shift can be determined by the amount of detuning.

Liao et al. [35] showed in stability analysis that the phase shift $\Delta\theta$ can be tuned over a 180° range. In the case of $\Phi = 0°$, $\Delta\theta$ lies in the range $-90° \leq \Delta\theta \leq +90°$.

The relationship between the successive phase shifts and the scan angle is given by

$$\Delta\theta = \frac{2\pi d}{\lambda} \sin\Psi \tag{5.27}$$

where d is the separation between the adjacent elements, λ is the wavelength, and Ψ is the scan angle measured from the broadside. We can increase the scan range by decreasing the element separation in the array. So, the beamwidth is broadened which can then be reduced by increasing the number of antenna array elements. There is no effect on the scan range by an increase in the number of array elements because $\Delta\theta$ is independent of the antenna array size. The angular resolution for the array is set by the stability and accuracy of the oscillators (VCOs) on the array periphery.

Experimental Results

Liao et al. [35, 36] reported a four-element and a six-element radiatively coupled active patch antenna arrays as shown Figures 5.12 and 5.13, respectively. A four-element array is manufactured on a 0.787-mm-thick Roger Duroid 5880 substrate with the relative permittivity of 2.2 and a NE3284A low-noise GaAs FETs. The width (W) and length (L) of each antenna element are shown in the figure as 4.56 mm and 11.88 mm, respectively.

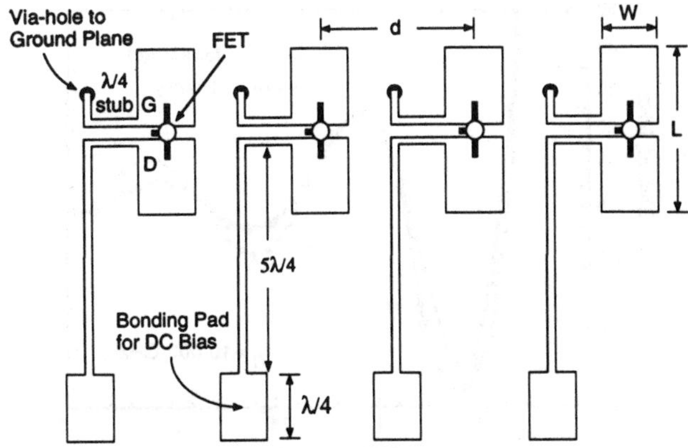

Figure 5.12 Sketch of a four-element beam-scanning FET array. (© 1993 IEEE.)

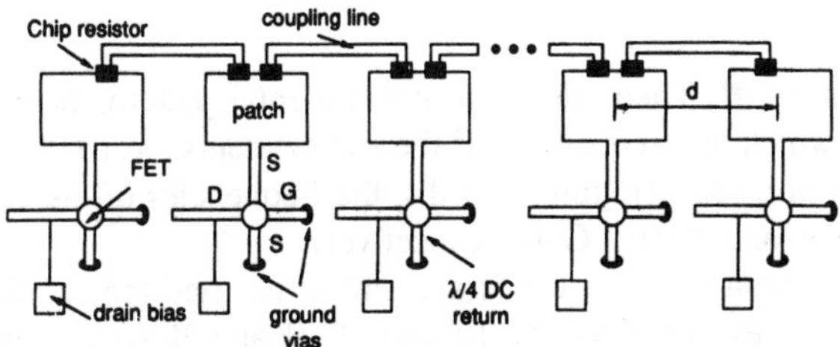

Figure 5.13 Sketch of a six-element beam-scanning FET array. (© 1994 IEEE.)

These values are selected empirically so that the oscillator is bias tunable over a range of frequencies centered about 10 GHz. In this design the FET source lead is grounded through a hole in the substrate. The frequency and tuning range is very sensitive to the inductance of the source lead. Liao et al. [35] designed the oscillator to operate at Vgs (*voltage between gate and source*) = 0, with a dc return path to ground provided by a quarter-wavelength shorted stub, connected to the circuit at a low impedance point. They used a typical bias network for a drain bias, which consists of a high-impedance 5 λ/4 line, followed by a quarter-wavelength impedance bonding pad. The coupling strength, coupling phase and the scanning range are determined by the oscillator separation. York et al. [35] used an imaging technique to characterize the coupling parameters and results as shown in Figure 5.14. The result shows that a center-to-center spacing of 0.86 λ is required to obtain $\Phi = 0°$, which limits the theoretical maximum scan to ±17° for the array that corresponds to an inner-element phase-shift range of $-80° < \Delta\theta < 66°$. Figure 5.15(a) shows the measured and theoretical broadside patterns that are obtained by setting all the oscillators' natural frequencies to 10 GHz. Figure 5.15(b) shows the measured and the-

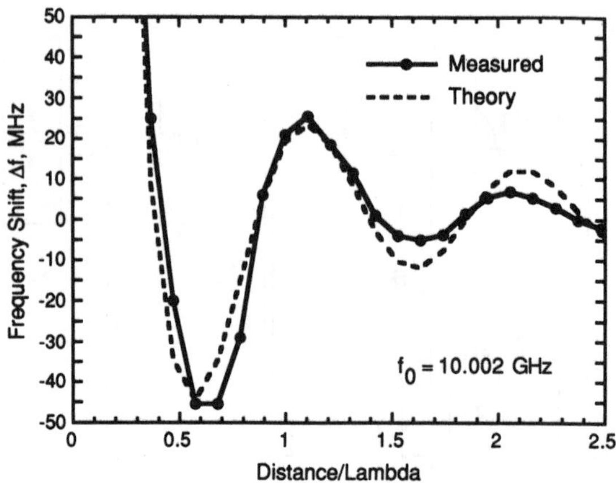

Figure 5.14 Plots of frequency shift versus distance in wavelength of a two-oscillator system. (© 1993 IEEE.)

oretical patterns at $-10°$. Theoretical and experimental patterns at $-15°$ are shown in Figure 5.15(c), in which the frequencies of the end elements are 10.0075 GHz and 9.9925 GHz. For the scan of ±12.5° (Figure 5.15(d)), the frequencies of the end elements in the array are 9.985 GHz and 10.015 GHz, respectively.

To extend the scan range of the array, Liao et al. used the oscillator design that is based on the techniques described by Johnson [37] and Kotzebue et al. [38]. For the oscillator design, Johnson reported that *S*-Parameter IS21I was the large signal and varied significantly from its small signal value. The large-signal values of *S*-Parameter IS21I are the point of maximum power-added efficiency. Optimum oscillator network design equations are reported in the 1975 Conference by Kotzebue et al. [38]. They computed lumped-element component values that would provide maximum oscillator power at the design frequency. Figure 5.13 shows a 4-GHz oscillator circuit and lumped-element values are obtained using the design approach of references [28–36]. The initial design of a six-element circuit is performed using a NE32184A GaAs MESFET and is constructed on a 31-mil-thick Rogers Duroid 5880 microwave substrate [39]. To reduce the size of the array, the fabrication is performed on 31-mil-thick Rogers Duroid 6010 [39] of the relative permittivity of 10.8. The length and width of the microstrip antenna element are 10.8 mm and 16.2 mm, respectively. The element provides a load impedance of about 400 Ω at resonance. A six-element beam-scanning array is shown in Figure 5.16. Each oscillator in the array contains the lumped-element circuit as shown in Figure 5.16. The coupling circuit is described by York et al. [33]. In the figure, each oscillator is coupled to its two nearest neighbors via a resistive load that is about one wavelength long with an element separation of 0.26 wavelength at 4 GHz. The coupling strength value of about 4 is obtained by using 50-Ω series resistors in the circuit.

The oscillators are coupled via the transmission line networks and the oscillation frequency changes from 4 GHz to 4.24 GHz. There is a change in the coupling phase from 0° to 27° due to the change in the oscillation frequency. The increment in frequency of oscillation changes the increase in electrical length of the antenna element separation from 0.26 λ to 0.29 λ. The maximum scan range for an antenna array with this element

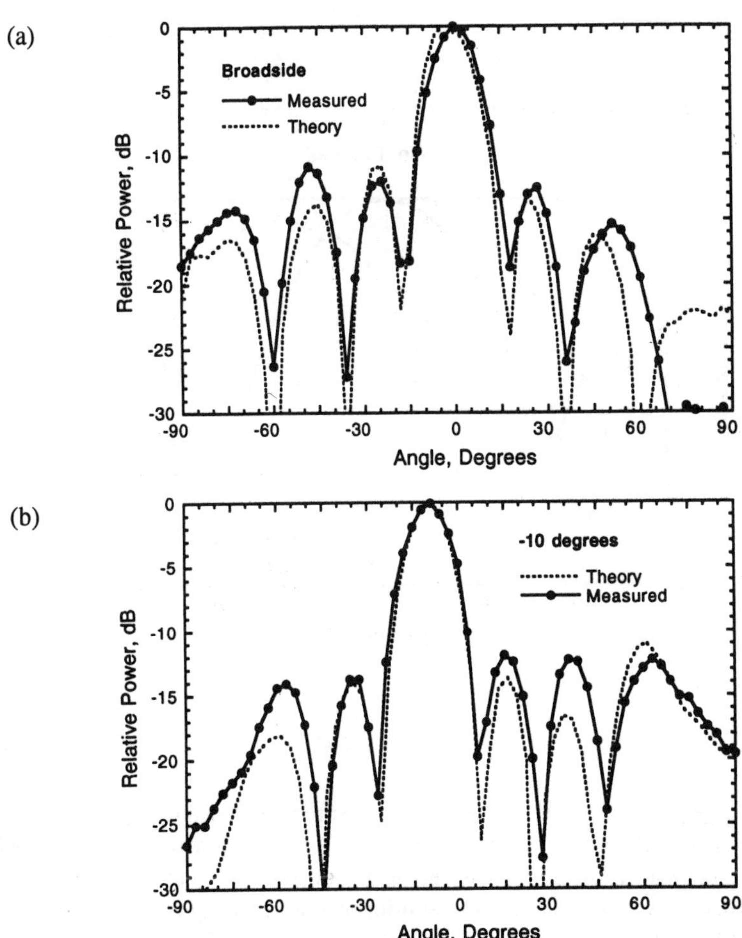

Figure 5.15 Experimental and theoretical broadside radiation patterns (a) when all oscillators have identical frequencies, (b) for $-10°$ scan, and (c) for $-15°$ scan, and (d) experimental broadside radiation patterns from $-15°$ to +12.5°. (© 1993 IEEE.)

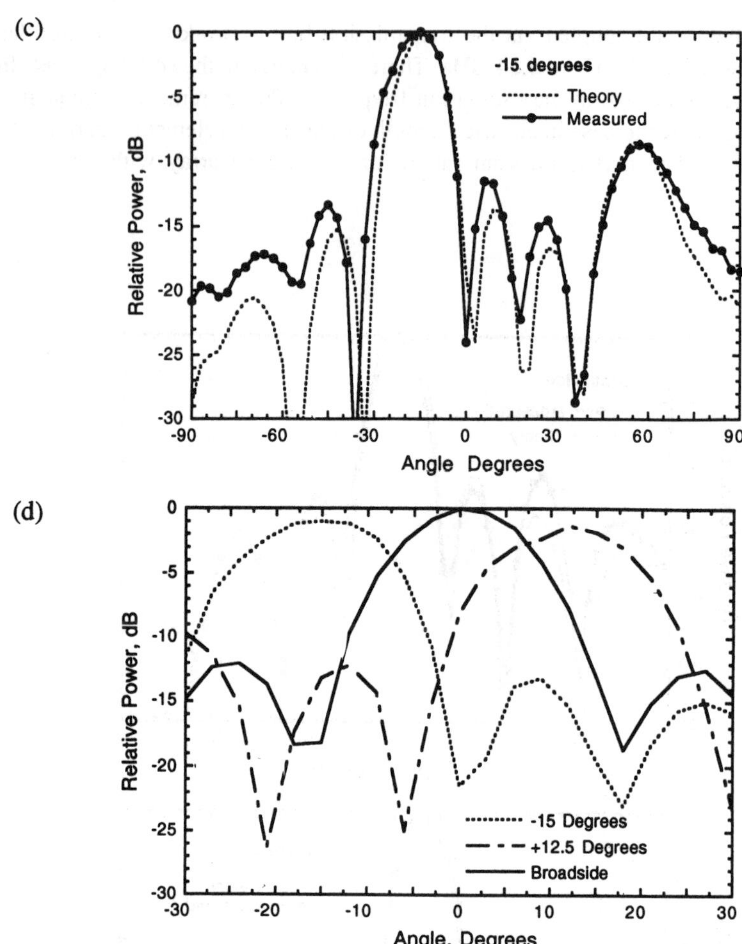

Figure 5.15 (continued)

separation translates to ±59°. Figure 5.17 shows measured beam-scanned patterns at +15°, +30°, and +40° when the free-running frequencies of the end elements are bias tuned as described by Liao et al. [36]. The measured scan range is −40° to +40°, whereas the theoretical prediction is ±59°. The main reason for this discrepancy is that the frequency-tuning range of the end oscillators is smaller than the locking bandwidth for the oscillators in this array. The progressive phase shift produced between the elements ($\Delta\theta$) is smaller than that required for maximum beam scan. The application of the wideband oscillator can provide the scan angle very close to the theoretical prediction.

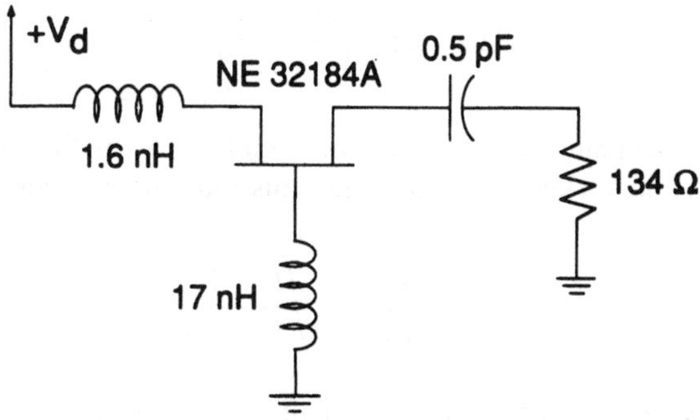

Figure 5.16 Design values of a 4-GHz oscillator circuit. (© 1994 IEEE.)

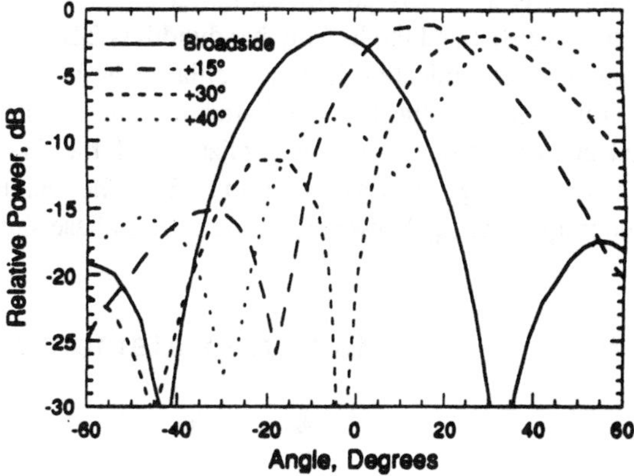

Figure 5.17 Measured radiation patterns for continuous scanning from broadside to +40°. (© 1994 IEEE.)

5.3.4 Unilateral Injection Locking

The unilateral injection-locking method provides more controlling capability than the interinjection-locking method [40–43]. In this method, the reverse injection locking is eliminated by using unilateral amplifiers between oscillators. This provides an advantage of using FETs in both amplifiers and oscillators so that this type of antenna array can be fabricated monolithically. The amplifiers in the circuit also control the locking bandwidths of the oscillators.

The phase controlling is performed by unilateral injection locking, which is based on Kurokawa's theory of injection locking [44]. Kurokawa described that when an oscillator is injection locked by an external signal, there is a phase difference Δ between the oscillation signal and injected signal. The phase difference $\Delta\phi$ is given by

$$\Delta\phi = \sin^{-1}\left(\frac{\omega_f - \omega_0}{\Delta\omega_m}\right) \tag{5.28}$$

where ω_f is the free-running frequency of the oscillator, ω_0 is the injected signal frequency, and $\Delta\omega_m$ is the locking range of a transmission type oscillator and is given by

$$\Delta\omega_m = \frac{\omega_0}{Q_{ext}} \frac{G_s}{G_p} \sqrt{\frac{P_i}{P_o}} \frac{1}{\sin\alpha} \tag{5.29}$$

where Q_{ext} is the external Q of the resonant circuit, α is the angle between the impedance locus and the device line, P_o is the output power, P_i is the injection power, G_s is the maximum stable gain of the two-port oscillator, and G_p is the square root of the output power ratio of the two-ports as described by Tajima et al. [45]. The maximum phase difference in this system is ±90°; and the locking bandwidth is $2\Delta\omega_m$. The design concept of an active array using unilateral injection-locked oscillators is shown in Figure 5.18. The reference signal ω_0 is used to lock the first oscillator; the phase shift is ϕ_1. The second oscillator has locked angular frequency ω_0 and phase 1 of the first oscillator, which introduces a phase shift of $\Delta\phi = \phi_2 - \phi_1$ as shown in (5.28). The locking frequency of all oscillators is the same frequency, ω_0. However, a progressive phase shift is created, which is described by

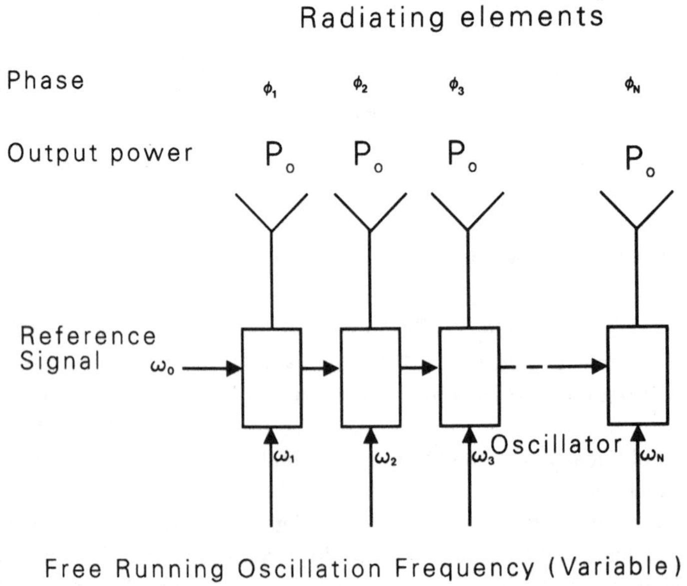

Figure 5.18 Concept of an active phased array using unilateral injection-locked oscillators.

$$\Delta\phi = \phi_n - \phi_{n-1} \tag{5.30}$$

This process is progressively established until all oscillators are locked. With the adjacent antenna radiating with phase difference $\Delta\phi$, the main beam can be scanned to an angle

$$\theta = \sin^{-1}\frac{\lambda\,\Delta\phi}{2\pi d} \tag{5.31}$$

where λ is the free-space wavelength and d is the antenna element spacing. The maximum beam scan is possible with the maximum phase shift $\Delta\phi_m = \pm 90°$. The maximum locking range, $\Delta\omega_m$, is proportional to the square root of injection power $P_i \cdot P_i$ can be varied, and $\Delta\omega_m$ is tuned by varying the amplifier gain and bias voltages.

Integrated Microwave Optical System

Recently, Chew et al. [46] described a conceptual configuration of an integrated microwave optical system as shown in Figure 5.19. In the figure, a multiple quantum-well (MQW) InGaAs-InGaAsP distributed feedback laser device is used for this work. The operating wavelength of the device is 1.55 μm; its threshold current and external quantum efficiency is 23 mA and 36.3 mW/A, respectively. When the laser is biased at 100 mA, the half-power point beamwidth is 10 GHz. Liau et al. [47] showed that the laser has a better relative intensity noise than the Fabry–Perot laser, which translates to a better signal-to-noise ratio performance. Higher power can be obtained by optimizing the coupling strength of the laser gratings.

Cox III et al. [48] showed a representative plot of the fiber-coupled optical power (p_o) versus laser current (i_L) for a directly modulated diode laser in Figure 5.20. Agrawal et al. [49] reported that p_o is related to i_L by slope efficiency

$$\eta_L(i_L) = \frac{dp_o}{di_L} \tag{5.32}$$

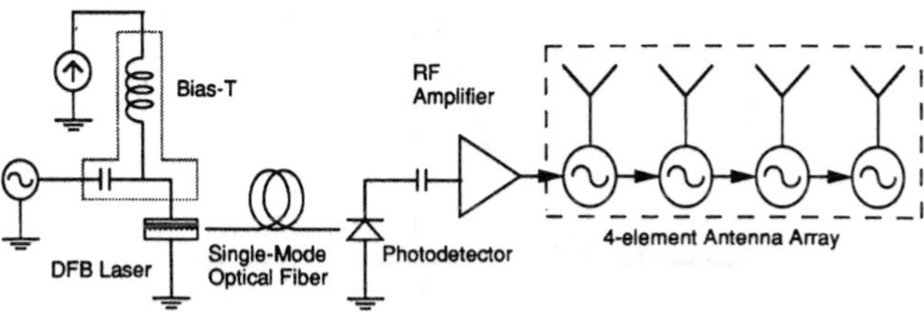

Figure 5.19 Block diagram of the integrated optical input/phased array antenna. (© 1994 IEEE.)

The total laser current is given by

$$i_L = I_L + i_l \tag{5.33}$$

In the case of efficient linear polarization, the bias current I_L must be greater than the lasting threshold current I_T. The modulation current i_l is superimposed on I_L, resulting in a total laser current as shown in the above equation.

When $i_l << I_L$,

$$\eta_L(i_L) \cong \eta_L(I_L) \equiv \eta_{LB} \tag{5.34}$$

where η_{LB} is the slope efficiency independent of i_l. For a small-signal model, a laser diode p_o is given by

$$P_o = \eta_{LB} i_l \tag{5.35}$$

In Figure 5.20, I_L is the operating point. The frequency response of the laser is dependent upon the resistance in series with the junction R_L and the bond wire inductance L_L. This is valid at frequencies well below the relaxation resonance as described by Agrawal et al. [49]. Figure 5.20 shows a small-signal lumped-element circuit model that contains a microwave source represented by an ideal voltage source (v_{in}), resistor R_{IN}, matching transformer, and diode laser [50]. The bond wire inductance L_L is not a laser parameter; it can be optimized independently. For $I_L > I_T$ the incremental voltage drop across the diode laser junction is negligible compared to the incremental voltage drop across R_L. Cox III et al. [48] deduced the following relationship between p_o and v_{in};

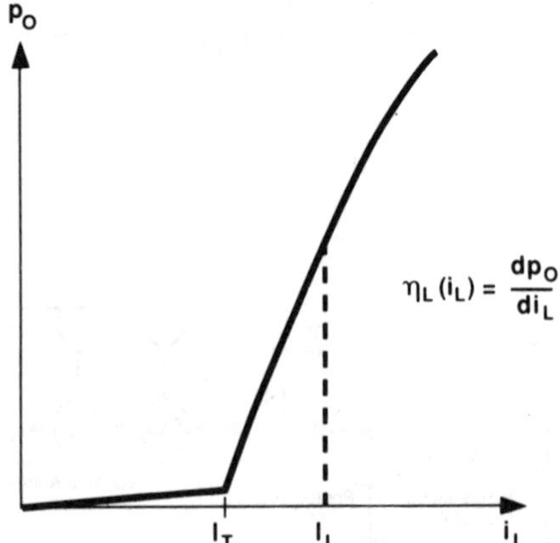

Figure 5.20 A typical plot of optical power versus laser current diagram for a laser diode [48].

$$p_o = \frac{\eta_{LB} N_L}{R_{T1}\,[(sN_L^2 L_L / R_{T1}) + 1]}\, v_{in} \tag{5.36}$$

where N_L is the turn ratio of the laser transformer, $R_{T1} = N_L^2 R_L + R_{IN}$ and s is the complex frequency represented by $\alpha + j\omega$; where $j\omega$ corresponds to the real and imaginary parts, respectively. To evaluate the magnitude of the frequency response, α is assumed to be zero. However, expressing equations in terms of the complex frequency makes it possible to evaluate the response of the device to more general inputs (a step input). The microwave power available from the source $P_{in,a}$ is

$$p_{in,a} = \frac{v_{in}^2}{4R_{IN}} \tag{5.37}$$

From (5.36) and (5.37), the incremental modulation efficiency for a laser diode is given by

$$\frac{p_o^2}{p_{in,a}} = \frac{4R_{IN}\,\eta_{LB}^2\,N_L^2}{R_{T1}^2\,[(sN_L^2\,L_L\,/\,R_{T1}) + 1]^2} \tag{5.38}$$

which can be rewritten, when R_L is matched to R_{IN} via $N_L(N_L^2 R_L = R_{IN})$, as

$$\left(\frac{p_o^2}{p_{in,a}}\right)_M = \frac{\eta_{LB}^2}{R_L\,[(sL_L\,/\,2R_L) + 1]^2} \tag{5.39}$$

where the subscript M denotes the matching condition. R_L is the only parameter for which a trade-off exists between gain and bandwidth. Schmidt [51] described some fundamental limits of the impedance matching versus bandwidth trade-offs. The shaded area in Figure 5.21 shows the incremental modulation efficiency versus bandwidth, which is calculated from (5.39). A dotted line in the figure denotes the modulation efficiency for the resistive match case, which is obtained from (5.39) by letting $N_L = 1$ and $R_L = R_{IN}$.

The p-i-n photodiode shown in Figure 5.19 can be represented by a lumped-element circuit model (Figure 5.22). This figure is represented by a p-i-n photodetector diode, a matching transformer, and an RF load. Analysis of this circuit provides the incremental modulation efficiency [48] as

$$\frac{p_{out}}{p_{od}^2} = \frac{\eta_D^2\,R_{OUT}\,N_D^2}{[sC_D\,(R_D + N_D^2\,R_{OUT}) + 1]^2} \tag{5.40}$$

where C_D is the diode junction shunt capacitance, R_D is the series resistance, R_{OUT} is the

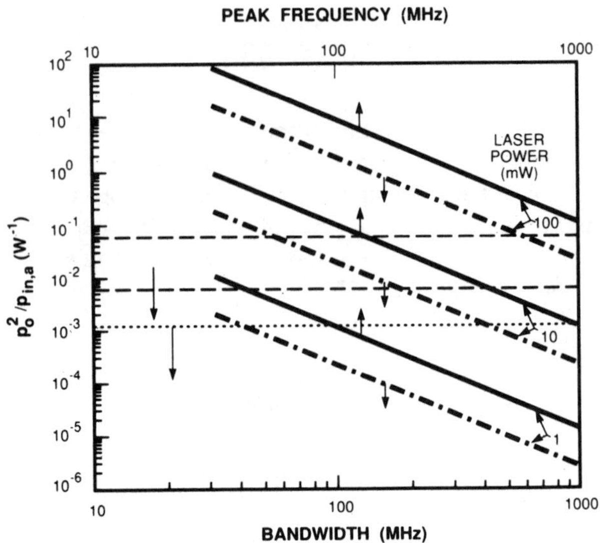

Figure 5.21 Plots of incremental modulation efficiency versus bandwidth [48].

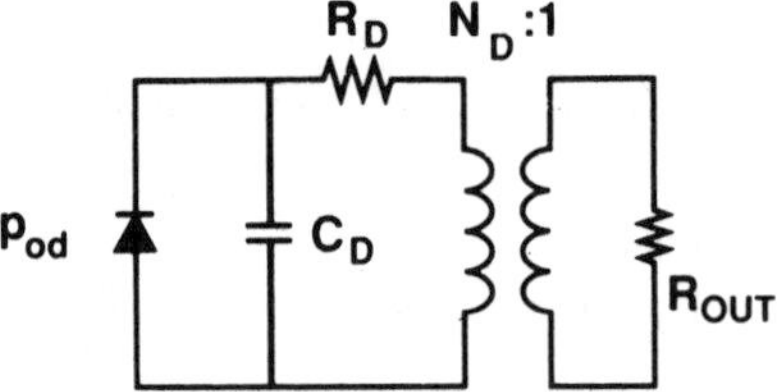

Figure 5.22 Small-signal lumped-element circuit model showing p-i-n photodiode, matching transformer, and RF load [48].

RF load, and N_D is the turn ratio of the photodiode transformer. The maximum modulation efficiency can be achieved by maximizing N_D, and the bandwidth will reduce as a result of increases in N_D. By assuming $\omega < [C_D(R_D + N_D^2 R_{out})]^{-1}$, (5.40) can be rewritten as

$$\frac{p_{out}}{p_{od}^2} = \eta_D^2 R_{out} N_D^2 \tag{5.41}$$

The link gain for our system (Figure 5.19) is given by

$$G = \frac{p_{out}}{p_{in,a}} = \frac{p_o^2}{p_{in,a}} t_{od}^2 \frac{p_{out}}{p_{od}^2} \tag{5.42a}$$

where p_{out} is the RF power delivered to R_{out} and t_{od} is the link optical transfer efficiency. Substituting (5.39) and (5.41) into (5.42) yields

$$G = \frac{1}{R_L} \eta_{LB}^2 \, t_{od}^2 \, \eta_D^2 \, R_{OUT} \, N_D^2 \tag{5.42b}$$

where G is independent of optical bias power and η_{LB}^2 has an upper bound determined by conservation of energy. The noise figure, *NF,* of the optical link is given by

$$NF = 10 \log 2 \left[1 + \frac{N_D^2 \, R_{IN}}{k \, T \, G} (n_1^2 + n_2^2) \right] \tag{5.43}$$

where T is the absolute temperature, k is Boltzmann's constant, n_1 is the relative noise, and n_2 is the shot noise at the p-i-n photodiode detector. Figures 5.23 and 5.24 show a plot of link gain versus laser power and noise figure versus laser power for η_{LB} = 0.14 W/A, R_L = 12.5 Ω, and t_{od} = 1, respectively.

A single-mode optical fiber is used and has a core diameter of 9 μm. The laser is biased at 45 mA to produce the average optical power of −3.5 dBm in the fiber. The current modulation index is estimated to be 6.8 percent. The signal from the fiber is passed through a high-speed p-i-n photodiode (HP8344OD) photodetector with a bandwidth of 34 GHz.

The overall insertion loss of the optical fiber link, including the detection and coupling loss, is measured to be −36 dB. The loss is due to the low quantum efficiency of the laser and impedance mismatch, which further reduces the link efficiency by a factor of approximately 0.34. Therefore, we can say that the link efficiency can be improved by improving the quantum efficiency of the laser, input impedance matching, optical-

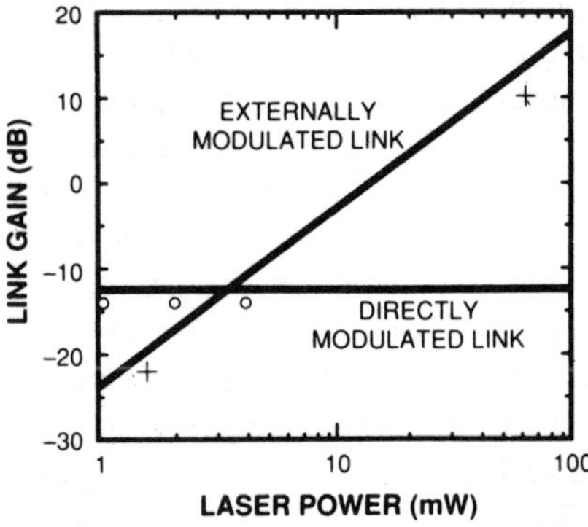

Figure 5.23 Experimental and theoretical plots of link gain versus laser power for directly modulated links (——— theoretical curve, o o o o experimental data). (© 1990 IEEE.)

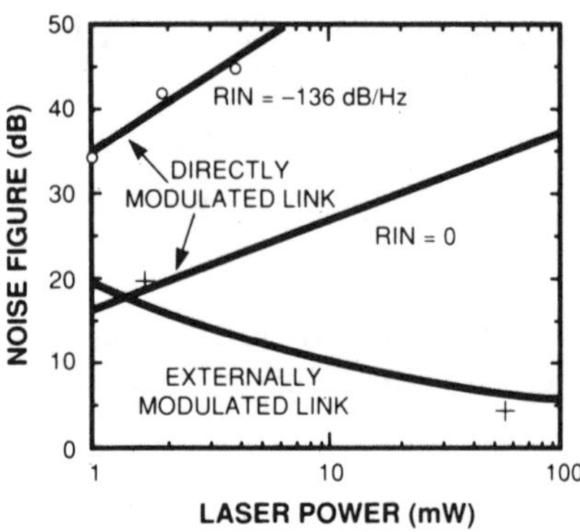

Figure 5.24 Plots of noise figure versus laser power for directly modulated links (——— theoretical curve, o o o o experimental data). (© 1990 IEEE.)

microwave conversion signal efficiency, and fiber coupling. The reconstructed (recovered) microwave signal passes through an RF (microwave) amplifier to boost the power level to 4 dBm for injection locking.

Experimental Results

The system setup of the integrated optical phased array antenna is shown in Figure 5.19. Lin et al. [40] and Chew [46] reported the circuit structures of an active two-element and a four-element active phased arrays at 6 GHz. Figure 5.25 shows a schematic diagram of the RF subcircuit of the active antenna. Two and four FETs are used in the cases of two- and four-element cases, respectively. Each active element antenna contains one FET and one terminal resistor. An amplifier is used between the active elements as shown in Figure 5.25. In these structures, rectangular microstrip antennas are used as radiators and resonators of the oscillators. A −20-dB directional coupler couples the input and output injection signals. In the case of a two-element structure, the unilateral amplifiers provides about 9-dB gain and −17-dB isolation to prevent reverse locking. Amplifiers with −25-dB isolation serve as active isolators between oscillators in the case of a four-element array. The gain of the amplifiers and the coupling level of the couplers are considered for optimum locking bandwidth. The whole circuit of both (two- and four-element) arrays uses a substrate with $\epsilon = 2.33$.

In the case of a two-element array, the drain bias of the FET in the second oscillator was changed to sweep the free-running frequency of the second oscillator through the

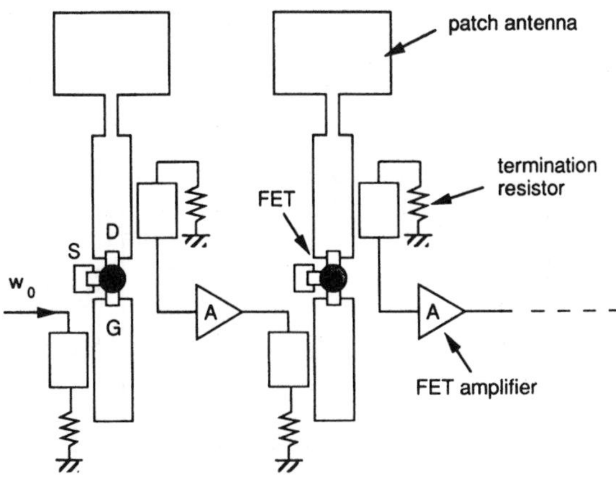

Figure 5.25 Schematic of a four-element active array for the unilateral injection locking. (© 1994 IEEE.)

injection-locking range. Lin et al. [40] used element spacing of 0.79 λ in the case of a two-element active array. The theoretical prediction of maximum beam scanning of 36.9° is calculated for the maximum phase shift of ±90° (180°). However, they observed a beam-scanning range of −27° to +6°. The maximum beam-scanning angle is 33°, which is very close to the theoretical prediction. Lin et al. [40] reported a beam-scanning range of −21° to +6° for the three-element array case. The drain biases of the FETs in the second and third oscillators were changed to sweep the free-running frequencies through the injection-locking range. The maximum beam-scanning angle is 27°, which is 6° less than the results of the two-element array. The measured antenna pattern for the four-element array is shown in Figure 5.26. The spacing between elements is 0.86 λ, and the calculated theoretical limit (maximum) of the scan angle is 33.8°. The measured maximum scan angle is about 21°, which is 12.8° less than the theoretical value. This discrepancy is due to the difficulty in achieving band-edge injection locking. Mutual coupling between elements provides inconsistency in phase shift along antenna elements and an introduction of additional phase shift in the circuit layout. Figure 5.26 shows the difference in power level of the main beam, which is caused mainly by the varied drain voltages of the oscillators.

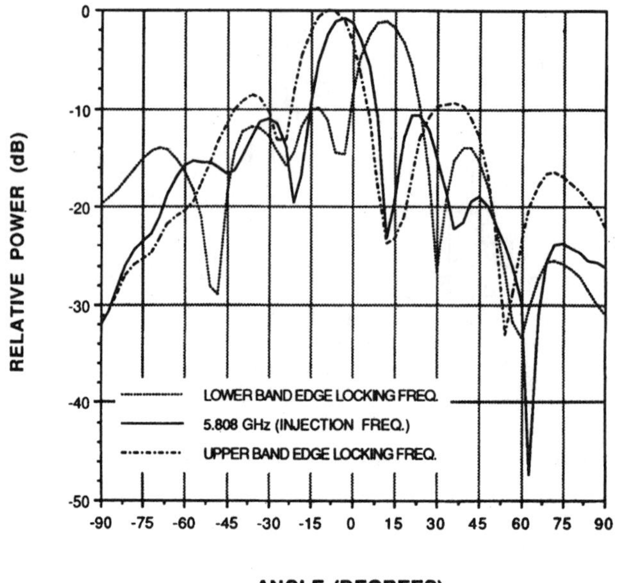

Figure 5.26 Measured scan patterns of the integrated four-element array using unilateral injection locking. (© 1994 IEEE.)

References

[1] Wolff, E. A., *Antenna Analysis,* New York: John Wiley and Sons, 1966.

[2] Hansen R. C., Ed., *Microwave Scanning Antennas,* New York: Academic Press, 1966.

[3] Jasik, H., Ed., *Antenna Engineering Handbook,* New York: McGraw-Hill, 1961.

[4] Rudge, A. W., K. Milne, A. D. Olver, and P. Knight, *The Handbook of Antenna Design, Volumes* 1 *and* 2, London, U.K.: Peter Peregrinus Ltd., 1986.

[5] Kumar, A., and H. D. Hristov, *Microwave Cavity Antennas,* Norwood, MA: Artech House, 1989.

[6] Daryoush, A., P. Herczfeld, V. Contarino, A. Rosen, Z. Turski, and P Wahi, "Optical Beam Control of mm-Wave Phased Array Antennas for Communications," *Microwave J.,* Vol. 30, No. 3, March 1987, pp. 97–104.

[7] Kam, M., J. Wilcox, and P. Herczfeld, "Design for Steering Accuracy in Antenna Arrays Using Shared Optical Phase Shifters," *IEEE Trans. Ant. Prop.,* Vol. AP-37, No. 9, September 1989, pp. 1102–1108.

[8] Daryoush, A. S. "Optical Synchronization of Millimeter-Wave Oscillator for Distributed Architecture," *IEEE Trans. Microwave Theory Tech.,* Vol. MTT-38, No. 5, May 1990, pp. 467–476.

[9] Daryoush A. S., T. Berceli, R. Saedi, P. R. Herczfeld, and A. Rosen, "Theory of Subharmonic Synchronization of Non-Linear Oscillator," *IEEE MTT-S Digest,* 1989, pp. 735–738.

[10] Adler, R., "A Study of Locking Phenomena in Oscillators," *Proc. IRE,* Vol. 34, 1946, pp. 351–357.

[11] Berceli, T., W. Jemison, P. Herczfeld, A. S. Daryoush, and A. Paolella, "A Double-Stage Injection Locked Oscillator for Optically Fed Phased Array Antenna," *IEEE Microwave Theory Tech.,* Vol. MTT-39, No. 2, February 1991, pp. 201–207.

[12] Zhang, X., X. Zhou, B. Aliener, and A. S., Daryoush, "A Study of Subharmonic Injection Locking for Local Oscillators," *IEEE Microwave Guided Wave Lett.,* Vol. 2, No. 3, March 1992, pp. 97–99.

[13] Daryoush, A. S., T. Berceli, R. Saedi, P. R. Herczfeld, and A. Rosen, ''Theory of Subharmonic Synchronization of Nonlinear Oscillators,'' *IEEE MTT-S Int. Microwave Symp. Digest,* 1989, pp. 735–738.

[14] Zhang, X., X. Zhou, and A. S. Daryoush, ''Characterizing the Noise Behavior of Subharmonically Locked Local Oscillators,'' *IEEE Trans. Microwave Theory Tech.,* Vol. MTT-40, No. 5, May, 1992, pp. 895–902.

[15] Schmideg, I, ''Harmonic Synchronization of Nonlinear Oscillators,'' *Proc. IEEE,* Vol. 59, 1971, pp. 1250–1251.

[16] Bussgang, J. L., L. Ehrman, and J. W. Graham, ''Analysis of Non-Linear System With Multiple Inputs,'' *Proc. IEEE,* Vol. 62, No. 8, August 1974, pp. 1085–1119.

[17] Lipsky, S. E., and A. S. Daryoush, ''Fiber-Optic Methods for Injection-Locked Oscillators,'' *Microwave J.,* Vol. 35, No. 1, January 1992, pp. 80–88.

[18] Herczfeld, P., A. S. Daryoush, A. Rosen, A. Sharma, and V. M. Contarino, ''Indirect Subharmonic Optical Locking of a Millimeter-Wave IMPATT Oscillator,'' *IEEE Trans. Microwave Theory Tech.,* Vol. MTT-34, 1986, pp. 1371–1375.

[19] Kumar, A. ''Optical Control of T/R Model Phased Array Antenna,'' Research Report AK-OPT-RES-88, AK Electromagnetique Inc., Quebec, Canada, 1988.

[20] Daryoush, A. S., ''Optical Synchronization of Millimeter-Wave Oscillator for distributed Architecture,'' Invited Paper, *IEEE Trans. Microwave Theory Tech.,* Vol. MTT-38, 1990, pp. 467–476.

[21] Sturzebecher, D. J., X. Zhou, X. Zhang, and, A. S. Daryoush, ''Design of Oscillators for Optically Controlled MMW Phased Arrays,'' *IEEE Int. Microwave Symp. Digest,* 1992, pp. 325–328.

[22] Esman, R., L. Goldberg, and J. Weller, ''Optical Phase Control of an Optically Injection-Locked FET Microwave Oscillators,'' *IEEE Trans. Microwave Theory Tech.,* Vol. MTT-37, No. 10, October 1989, pp. 1512–1518.

[23] Daryoush, A. S., M. Francisco, R. Saedi, D. Polifko, and R. Kunath, ''Phase Control of Optically Injection Locked Oscillators for Phased Arrays,'' *IEEE Int. Microwave Symp. Digest,* May 1990, pp. 1247–1250.

[24] Zhang, X., and A. S. Daryoush, ''Full 360 deg. Phase Shifting of Injection-Locked Oscillators,'' *IEEE Microwave Guided Wave Lett.,* Vol. 3, No. 1, January 1993, pp. 14–16.

[25] Kumar, A., ''Construction of a 360 deg. Phase Shifting Injection-locked Oscillators for Large Phased Array—Space Applications,'' Research Report, AK-OPT-SPA-94, AK Electromagnetique Inc., 1994.

[26] Stephan, K. D., ''Inter-Injection-Locked Oscillators for Power Combining and Phased Arrays,'' *IEEE Trans. Microwave Theory Tech.,* Vol. MTT- 34, No. 10, October 1986, pp. 1017–1025.

[27] Stephan, K. D., and S. L. Young, ''Mode Stability of Radiation-Coupled Interinjection-Locked Oscillators for Integrated Phased Arrays,'' *IEEE Trans. Microwave Theory Tech.,* Vol. MTT-36, No. 5, May 1988, pp. 921–924.

[28] Stephan, K. D., and W. A. Morgan, ''Analysis of Inter-Injection-Locked Oscillators for Integrated Phased Arrays,'' *IEEE Trans. Ant. Prop.,* Vol. AP-35, No. 7, July 1987, pp. 771–781.

[29] Rutledge, D. B., Z. B. Popovic, R. M. Weikle, M. Kim, K. A. Potter, R. A. York, and R. C. Compton, ''Quasi-Optical Power Combining Arrays,'' Invited Paper, *IEEE MTT-S Int. Microwave Symp. Digest,* Dallas, Texas, May 1990.

[30] York, R. A., and R. C. Compton, ''Quasi-Optical Power Combining Using Mutually Synchronized Oscillator Arrays,'' *IEEE Trans. Microwave Theory Tech.,* Vol. MTT-39, No. 6, June 1991, pp. 1000–1009.

[31] York, R. A., R. M. Martinez, and R. C. Compton, ''Hybrid Transistor and Patch Antenna Element for Array Applications,'' *Electron. Lett.,* Vol. 26, March 1990., pp. 494–495.

[32] York, R. A., and R. C. Compton, ''Experimental Observation and Simulation of Mode-Locking in Coupled-Oscillator Arrays,'' *J. Appl. Phys.,* Vol. 71, No. 6, March 1992, pp. 2959–2965.

[33] York, R. A., and R. C. Compton, ''Measurement and Modelling of Radiative Coupling in Oscillator Arrays,'' *IEEE Trans. Microwave Theory Tech.,* Vol. MTT-41, No. 3, March 1993 pp. 438–444.

[34] York R. A., ''Nonlinear Analysis of Phase Relationships in Quasi-Optical Oscillator Arrays,'' *IEEE Trans. Microwave Theory Tech.,* Vol. MTT-41, No. 10, October 1993, pp. 1799–1809.

[35] Liao, P., and R. A. York, ''A New Phase-Shifterless Beam Scanning Technique Using Arrays of Coupled Oscillators,'' *IEEE Trans, Microwave Theory Tech.,* Vol. MTT- 41, No. 10. October 1993, pp. 1810–1815.

[36] Liao, P., and R. A. York, ''A Six-Element Beam-Scanning Array,'' *IEEE Microwave and Guided Wave Lett.,* Vol. 4, No. 1, January 1994, pp. 20–22.

[37] Johnson, K. M., ''Large Signal GaAs MESFET Oscillator Design,'' *IEEE Trans. Microwave Theory Tech.,* Vol. MTT-27, No. 3, March 1979, pp. 217–227.

[38] Kotzebue, K. L., and W. J. Parrish, ''The Use of Large Signal S-Parameters in Microwave Oscillator Design,'' *Proc. IEEE Int. Microwave Symp. Circuit and Systems,* 1975.

[39] Rogers Corporation Technical Data on Duroid 5880 and Duroid 6010, Texas, U.S.A.

[40] Lin, J., S. Kawasaki, and T. Itoh, ''Optical Control of MESFET's for Active Filter and Active Antenna,'' *Proc. Seventh Int. MIOP'93 Conf.,* Sindelfingen, Germany, May 1993, pp. 348–352.

[41] Kawasaki, S., and T. Itoh, ''Optical Control of 2-Element CPW Active Integrated Antenna with Strong Coupling,'' *IEEE AP-S Int. Symp. Digest,* June 1993, pp. 1616–1619.

[42] Berceli, T., ''Combined Lightweight-Microwave Technologies for Advanced Communications Systems,'' *Fourth Int. Symp. Recent Advances in Microwave Technology,* Wiley Eastern Ltd., New Delhi, India, December 1993, pp. 755–758.

[43] Kumar, A., Discussion on Injection Locking Methods with T. Berceli and T. Itoh at the 4th Int. Symp. on Recent Advances in Microwave Tech., New Delhi, India, December 1993.

[44] Kurokawa, K., ''Injection Locking of Microwave Solid-State Oscillators,'' *Proc. IEEE,* Vol. 61, No. 10, October 1973.

[45] Tajima, Y., and K. Mishima, ''Transmission-Type Injection Locking of GaAs Schottky-Barrier FET Oscillators,'' *IEEE Trans. Microwave Theory Tech.,* Vol MTT-27, No. 5, May 1979, pp. 386–391.

[46] Chew, S. K., T. K. Tong, M. C. Wu, and T. Itoh, ''An Active Phased Array With Optical Input and Beam Scanning Capability,'' *IEEE Microwave Guided Wave Lett.,* Vol. 4, No. 10, October 1994, pp. 347–349.

[47] Lau, K. Y., C. M. Gee, T. R. Chen, N. Bar-Chaim, and I. Urym, ''Signal-Induced Noise in Fiber-Optic Links Using Directly Modulated Fabry–Perot and Distributed-Feedback Laser Diodes,'' *J. Lightwave Tech.,* Vol. 11, 1993, pp. 1216–1225.

[48] Cox III, C. H., G. E. Betts, and L. M. Johnson, ''An Analytic and Experimental Comparison of Direct and External Modulation in Analog Fiber-Optic Links,'' *IEEE Trans. Microwave Theory Tech.,* Vol. MTT-38, No. 5, May 1990, pp. 501–509.

[49] Agrawal G. P., and N. K. Dutta, *Long Wavelength Semiconductor Lasers,* New York: Van Nostrand Reinhold, 1986.

[50] Liau, Z. L., and J. N. Walpole, ''A Novel Technique for GaInAsP/InP Burried Heterostructure Laser Fabrication,'' *Appl. Phys. Lett.,* Vol. 40, 1982, pp. 568–570.

[51] Schmidt, R. V., ''Integrated Optics Switches and Modulators,'' *Integrated Optics,* New York: Plenum, 1987.

Chapter 6

Hardware-Compressive Fiber Optical Delay Lines for Steering of Phased Array Antennas

6.1 INTRODUCTION

High-performance communications and radars require the use of *phased array antennas* (PAAs) composed of elemental monolithic *transmit/receive* (T/R) modules. The advantages of these antennas are steering without physical movement, accurate beam pointing and increased scan flexibility in one and two dimensions, reduced power consumption and weight, precise elemental phase, and amplitude control to obtain low spatial sidelobes. The actual number of T/R modules depends on the system mission as well as its operating frequency. To satisfy the wide *bandwidth* (BW) requirements of future PAAs, *true time delay* (TTD) steering techniques must be implemented so that an efficient elemental vector summation/distribution independent of frequency can be obtained. A fiber optic delay line system is designed to generate the delays required for a TTD-based PAA. This chapter describes hardware-compressive fiber optic delay lines and their applications in steering of phased array antennas.

6.2 DEVELOPMENT OF BINARY OPTICAL DELAY LINES

A three-bit optical time shifter that consists of laser diodes, bias switchings, eight optical fiber delay lines, a commercial star coupler, and a photodetector is shown in Figure 6.1. The delay lines provide 2^3 discrete delay increments for achieving three bits of resolution. The minimum differential time delay, Δt, corresponds to the least significant bit of the three-bit time shifter. At the time of antenna steering, one delay line is selected from each of the four such modules to provide time delays for four antenna subarrays. In the figure, each delay line of the module is connected by a 1.3-μm, p-substrate buried crescent laser diode. Ng et al. [1–3] biased the laser diode so that its resonance frequency was at about 7 GHz to achieve a good signal/noise ratio at 9 GHz. The *relative intensity noise* (RIN)

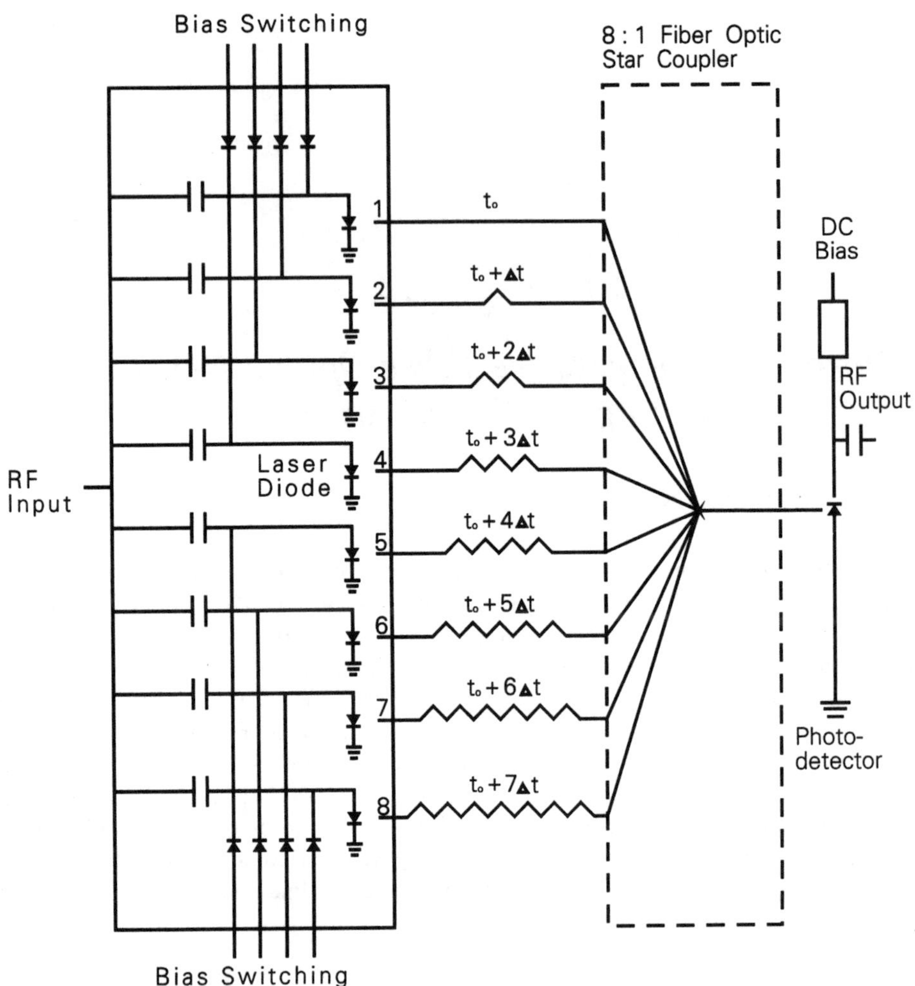

Figure 6.1 Schematic of 3-bit fiber-optic delay lines providing time delays from t_0 to t_0 + Δt as numbered 1 to 8. (© 1991 IEEE.)

was about 144 dB/Hz at 2 GHz and 9 GHz for the bias current I = 80 mA. The RF/microwave is directly modulated to the laser diodes. A specific delay increment 0, Δt, . . . , 7 Δt for channels 1 to 8 is introduced for each module by switching on the bias of the laser diode pig-tailed to the appropriate length of a multimode optical fiber. At the multimode star combiner (1 to 8), an optical loss of 10 dB was noticed and an average optical power of about 0.8 mW was found at the photodetector end when the LDs were biased at about 85 mA. Power amplifiers were used to boost the RF/microwave signal back to 10 dB≠m for a gain of unity during steering of the phased array.

The difference in time delay Δt_d between a particular line and shortest line is given by

$$\Delta t_d = (\Delta\phi/\Delta f)/360 \tag{6.1}$$

where Δf is the corresponding frequency sweep and the differential phase $\Delta\phi = (\phi_i - \phi_d$, $i = 1, 2, \ldots, 7)$. Here ϕ_d corresponds to Δt_d. The smallest differential delay increment, Δt, in each module is given by

$$\Delta t = L' (\sin \Theta_{max})/7c \tag{6.2}$$

where L' denotes the separation between the centers of the first and fourth subarrays, c is the velocity of light, and Θ_{max} is the maximum steering angle of the array.

The differential RF/microwave insertion phase between the input and outputs ports of delay lines #3 ($\Delta t_d = 2\,\Delta t$) and #5 ($\Delta t_d = 4\,\Delta t$) are shown in Figure 6.2. Figure 6.3 shows the differential insertion phase of delay lines #4 ($\Delta t_d = 3\,\Delta t$) and #7 ($\Delta t_d = 6\,\Delta t$). In Figures

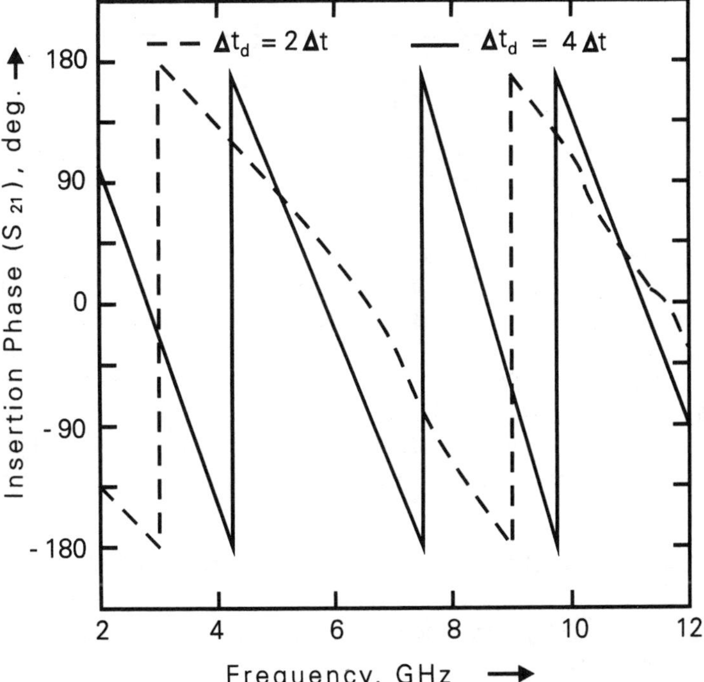

Figure 6.2 Differential RF/microwave insertion phase between the input and output ports of delay lines #3 ($\Delta t_d = 2\,\Delta t$) and #5 ($\Delta t_d = 4\,\Delta t$). (© 1991 IEEE.)

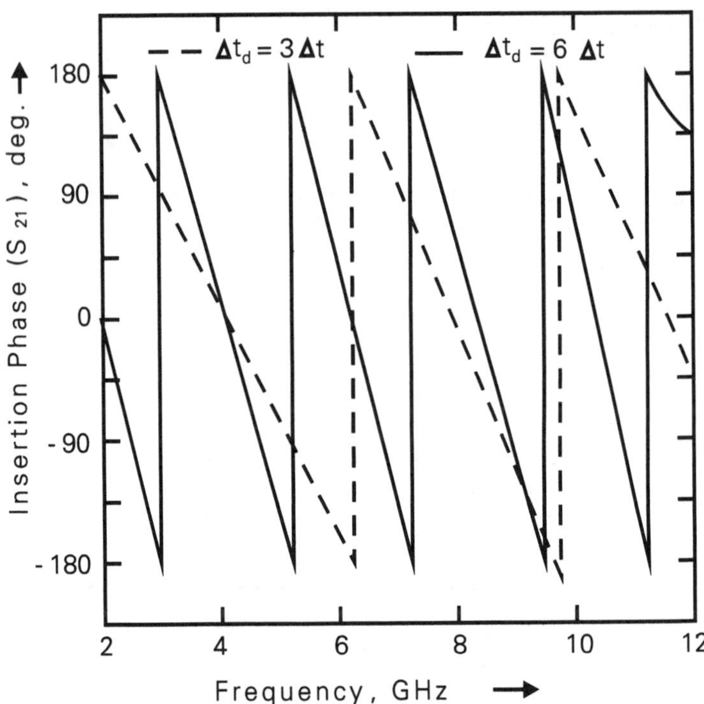

Figure 6.3 Differential RF/microwave insertion phase between the input and output ports of delay lines #4 ($\Delta t_d = 3\ \Delta t$) and #7 ($\Delta t_d = 6\ \Delta t$). (© 1991 IEEE.)

6.2 and 6.3, the input and output ports' delay lines have a ratio of 2 in the same module. In the figure the slopes of delay lines #5 and #7 are twice that of delay lines #3 and #4, respectively. The differential phases ($\Delta\phi$) of the remaining delay lines #2, #6, #8 in the same module are plotted in Figures 6.4 and 6.5. From Figures 6.4 and 6.5, we can conclude the following: (1) The differential phase linearity between the lines improves with a larger disparity between fiber lengths. (2) Plots indicate that while the *S/N* ratio of the differential phase is slightly degraded by the high-frequency roll-off in the laser modulation response, excellent phase linearity is maintained from 2 GHz to 12 GHz.

6.3 DELAY-COMPRESSIVE FIBER OPTICAL DELAY LINES

There are two types of delay lines that achieve significant hardware-compression. These delay lines are as follows.

- The *square root cascade delay line* (SRODEL) is a dual-cascaded-segment delay line where the desired delay segment is selected from a set of the square root of the

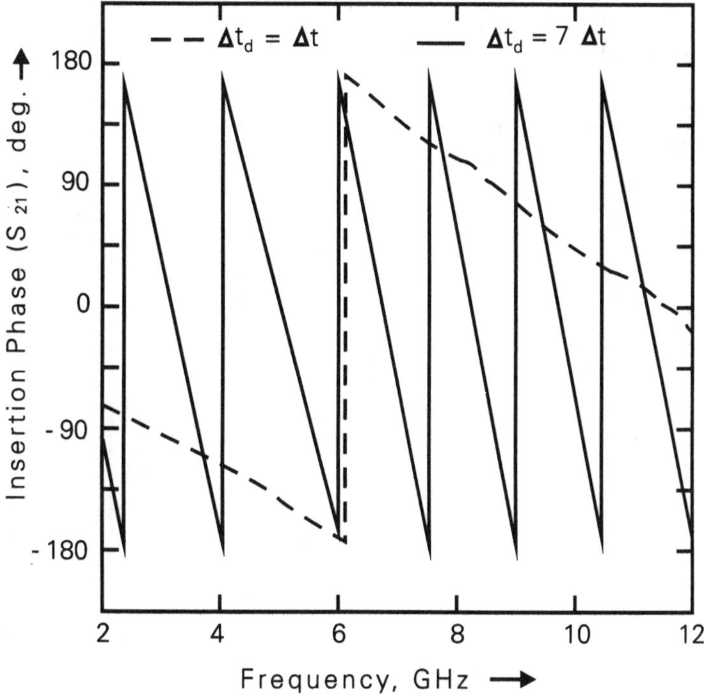

Figure 6.4 Differential RF/microwave insertion phase of delay lines #2 ($\Delta t_d = \Delta t$) and delay lines #8 ($\Delta t_d = 7\ \Delta t$). (© 1991 IEEE.)

number of bits. In this architecture, the fiber complexities, the switch complexities, and the dynamic range loss are related to the square root of the number of bits.

- The *binary fiber optical delay line* (BIFODEL) is a delay line where the desired delay segment is selected from a set of the logarithm base 2 of the number of bits. In this case, the optical signal is optionally routed through *N* fiber segments whose lengths increase successively by a factor of 2. The fiber complexities, the switch complexities, and the dynamic range loss are related to the logarithm (base 2) of the number of bits.

6.3.1 Square Root Cascade Delay Line (SRODEL)

A high-speed optical time switch with integrated optics was reported by Kondo et al. [4] in 1983. He developed a single polarization fiber delay line and Ti:$LiNbO_3$ (titanium diffused lithium niobate) integrated optical circuits.

Soref [5] extended Kondo's work to a combination of fiber-and-integrated optics two-stage design. In Figure 6.6, Soref described a voltage-controlled 64-delay device

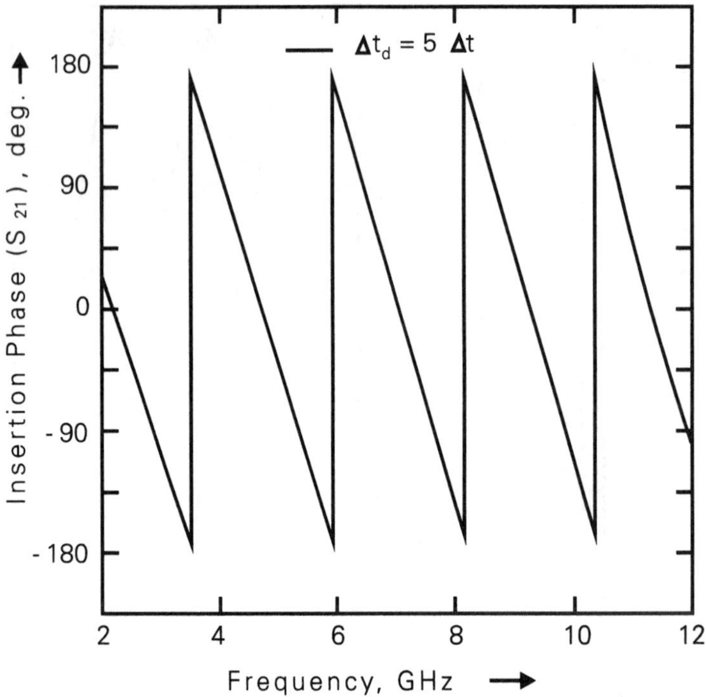

Figure 6.5 Differential RF/microwave insertion phase of delay lines #6 ($\Delta t_d = 5\ \Delta t$). (© 1991 IEEE.)

consisting of two cascaded transmissive stages located on the same integrated-optical wafer. The wafer contains two 1 × 8 electro-optical switching networks and two 8 × 1 electro-optical switching networks symbolized in the diagram by the branching arrays of 1 × 2 elements. Two groups of fibers, eight strands in each, are doubly coupled to the wafer. Corresponding switch elements are addressed electrically so that one fiber delay path in the first stage is selected along with one fiber path in the second stage. The main attraction of this architecture is that the fiber delays are additive and $N \times M$ difference delays are calculated from $N + M$ fibers. In the architecture, Soref [5] showed that the lengths of the first-stage fibers are $L_{11} = L_1$, $L_{12} = L_1 + \Delta L_1$, $L_{13} = L_1 + 2\Delta L_1$, etc. and that the second-stage fibers have lengths $L_{21} = L_2$, $L_{22} = L_2 + \Delta L_2$, $L_{23} = L_2 + 2\Delta L_2$, etc. The time delay ($t_{ij}$) is given by

$$t_{ij} = (L_{1i} + L_{2j})/v \tag{6.3}$$

where v is the group velocity of the fiber. The shortest delay increments in the first and second stages are, respectively, given by

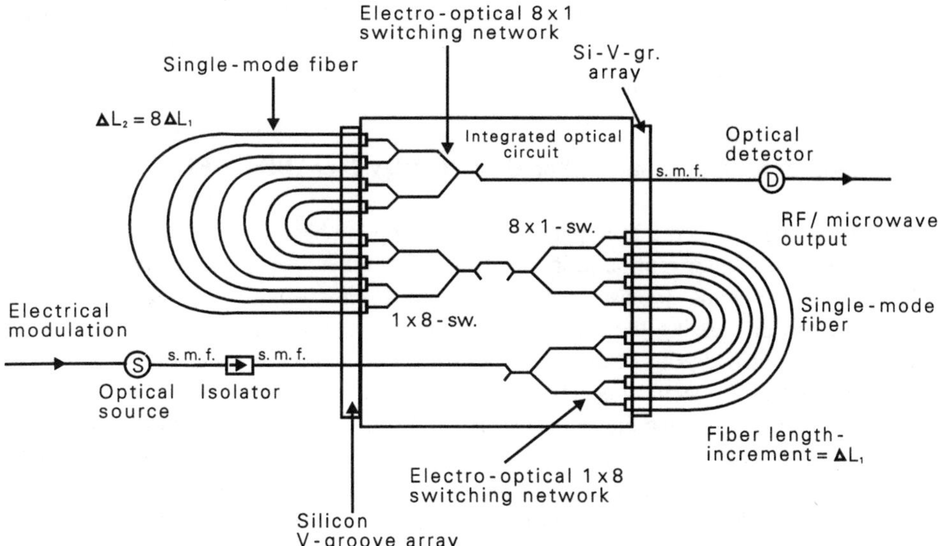

Figure 6.6 Programmable 64-delay device using T + T stages. (© 1984 OSA.)

$$\Delta t_1 = \Delta L_1/v \tag{6.4a}$$

$$\Delta t_2 = \Delta L_2/v \tag{6.4b}$$

When there are N fibers in each stage, ΔL_2 is given by

$$\Delta L_2 = N\,\Delta L_1 \tag{6.4c}$$

Equation (6.2c) provides a set of N^2 equal delay increments in a device that can produce any of the N^2 time steps Δt, $2^2\,\Delta t$, $3^2\,\Delta t$, . . . , $N^2\,\Delta t$.

In this system, the number of fibers, M_f, is used as delay paths, and the number of optical switches, M_s, is used to route the optical signal into various delay paths. The loss due to switch crosstalk is denoted by the DR_{loss}. The delay line reconfiguration time T_r is determined by the switch T_s. This architecture is a dual-cascaded-segment delay line where the desired delay segment is selected from a set of $(R)^{1/2}$ segments. Selection of the appropriate delay path is made using a tree-type structure of 1×2 optical switches at the input of each set. The various performance parameters for the SRODEL architecture are given as [6]

$$M_f = 2\,(R)^{1/2} \tag{6.5}$$

$$M_s = 2\,[(R)^{1/2} - 1] \tag{6.6}$$

$$DR_{\text{loss}}\ (\text{dB}) = 10\ \log_{10}[2\ \log_2(R)^{1/2}] \tag{6.7}$$

where R is the number of bits.

6.3.2 Background of Binary Fiber Optical Delay Line (BIFODEL)

In 1975, Levine [7] filed a U.S. patent on a fiber optic phased array antenna system for RF transmission. The invention in its most basic form involves apparatus for producing a phase-shifted RF signal for each antenna element by imposing the signal on an optical frequency carrier, passing the modulating carrier through an optical delay line and demodulating to provide the desired phase-shifted signal. Figure 6.7 shows a schematic block diagram of a system in accordance with Levine's invention, in which phase delay switching is effected after the optical-to-RF transducers. An optical carrier modulator modulates a microwave/RF signal with an optical signal and the modulated signal is fed to a multichannel fiber optic delay line via optical switches. The delay line signal is fed to an optical-to-RF transducer, which is controlled by a phasing programmer. The output

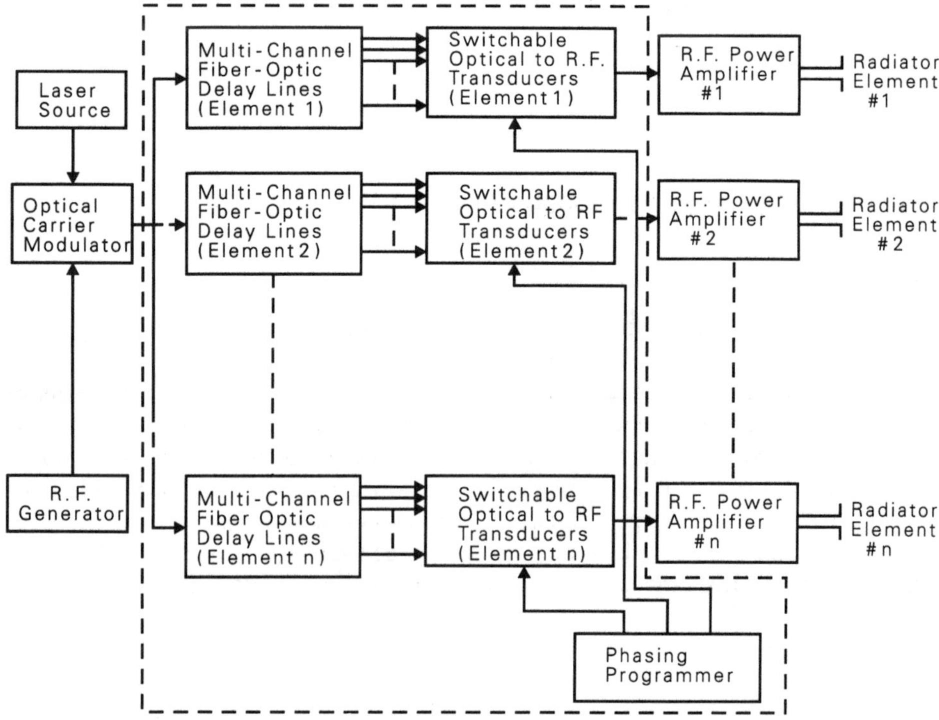

Figure 6.7 A block diagram of the optical fiber switching for controlling the phase delay [7].

from the switchable transducer is fed to an RF power amplifier and a radiating element. Each radiating element contains optical switches, multichannel fiber optic delay lines, a switchable optical-to-RF transducer including a phasing programmer, and an RF amplifier. In the figure, dotted lines show a block that contains multichannel fiber optic delay lines, switchable optical-to-RF transducers, and a phasing programmer for elements 1, . . . , *n*. This block can be replaced by a digitally controlled section of the fiber optic delay line for the antenna elements. Figure 6.8 shows a 5-bit digitally controlled section of the fiber optic line for one element. This digitally controlled section can be repeated for a number of elements (1, . . . , *n*) to replace the dotted line section of Figure 6.7. In a practical arrangement, the digital control signals employ the required number of bits to represent the full range of discrete beam positions that is consistent with the predetermined scan or beam-pointing granularity desired. The digital input signal at the digital control signal box includes the most significant digit, applied at A, the next most significant at B, and so on through C and D, down to the least significant digit applied at E. Each switch either diverts the optical signal through the length of fiber optic delay line that follows it or through a path that constitutes an optical ''short circuit'' to the switch. Thus for a 1 condition of the

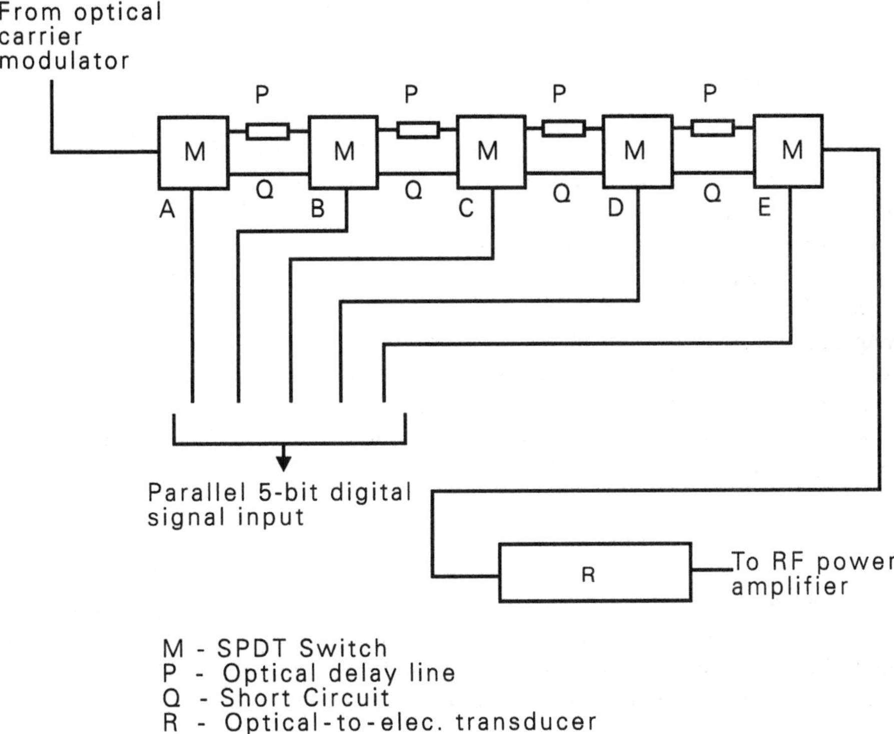

Figure 6.8 A block diagram of the optical fiber switching for control of the phase delay [7].

most significant digit lead A the output of the parallel 5-bit digital control signal box is diverted or fed to the optical switch 2 through the optical delay line 1, which has a predetermined delay that is consistent with the value of this most significant digit. In the 0 condition at A, the signal from the optical switch 1 reaches the optical switch 2 via the optical "short circuit." The identical process applies to each of the remaining digits in the control code word and, hence, the output on the optical-to-electrical transducer (photodiode) is delayed in accordance with the sum of the values of the 1 digits applied at the parallel 5-bit digital control signal box. For the same antenna parameters and other design considerations, the range of RF/microwave phase delays obtained at the output of the optical photodiode in response to the delayed light energy signal on the output of the optical switches is substantially identical to that provided at each given RF/microwave power amplifier that feeds a corresponding radiator in Figure 6.7. The output of the photodiode is applied to the corresponding RF/microwave power amplifier and the antenna element. It is realized that the rate of scan or beam positioning is an independent variable determined by the speed of operation of the phasing programmer in the case of optical phase control technique as shown in Figure 6.7. In the case of a parallel 5-bit digital (binary) signal control, the rate of scan or beam positioning is determined by a digital computer that supplies the control code at the binary control box.

6.4 BIFODEL ARCHITECTURE

The programmable BIFODEL has been reported by many researchers [7–9] for the steering of phased array antennas. Taylor [10], Levine [7], Lagerstroem et al. [11], Goutzoulis et al. [12], and Ng et al. [13] described a schematic of the BIFODEL architecture as shown in Figure 6.9. In the figure, it is shown that an optical signal is modulated by the microwave signal; this signal is routed through N fiber segments whose lengths are arranged so that the corresponding delays increase successively by a power of 2. The required fiber segments are addressed using a set of N 2×2 optical switches. Each optical switch allows the signal either to connect or to bypass a specific fiber segment. The connect and bypass states of the switch are denoted by $b_i = 1$ and $b_i = 0$, respectively. A delay T can be inserted in the circuit by an increment of ΔT by selecting the state $b_i = 1$ or 0 of the switches. The maximum value of the delay can be given by

$$T_{\max} = (2^N - 1)\Delta T \tag{6.8}$$

The modulated optical signals are routed through the appropriate fiber segments to produce delay. The number of fiber segments N ($N = M$) produces R delay. The number of optical fiber segments is also called the optical switch complexities (M_f, M_s). The optical switch complexities can be calculated from the expression

$$M_f = M_s = \log_2 R \tag{6.9}$$

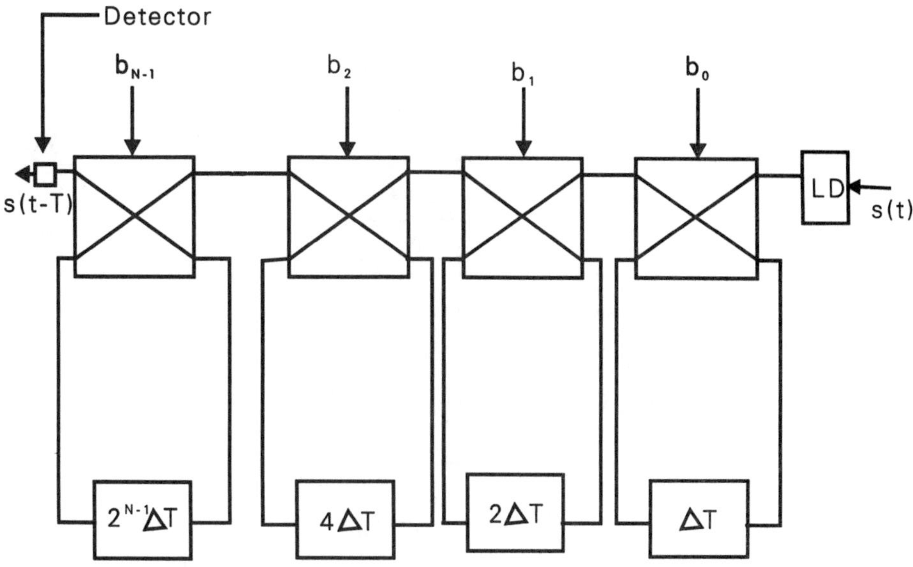

Figure 6.9 Programmable binary fiber optic delay line architecture. (© 1989 SPIE.)

where R is a delay resolution in bits.

The BIFODEL can be implemented with a combination of fiber and/or free-space delays and offer log-level compressive fiber switch complexities. The dynamic range loss due to optical switch crosstalk loss is known as DR_{loss}, which is given by

$$DR_{\text{loss}} = 10 \log_{10}(\log_2 R) \tag{6.10}$$

Ng et al. [13] reported a binary cascade delay line for the phased array antenna. Figure 6.10 shows a diagram of a binary cascade that enables 2^N time delay to be generated with N cascade time-shifters, each consisting of two delay lines. The RF/microwave signal is fed to a *laser diode* (LD), and the modulated optical signal is divided into two parts as shown in the figure. The signals one and two pass through the reference and delay arm, respectively. The reference signal does not provide any delay and it is called the "no delay signal," and the signal that passes through the delay arm provides a delay and is called the "delay signal." It is shown in the Figure 6.10 that the delay arm in each cascade stage is twice as long as that of the preceding stage. The switch in these cascade stages is controlled by the logic level of specific bit. 2^N delay times can be generated with $2 \times N$ physically distinct delay segments by switching the RF-modulated optical beam into either the delay signal or no delay signal arm in each stage. We can insert the required time delay (T) by selcting the appropriate switch position. The maximum time delay can be achieved by this architecture as reported by Goutzoulis et al. [14]. The cascade architecture reduces

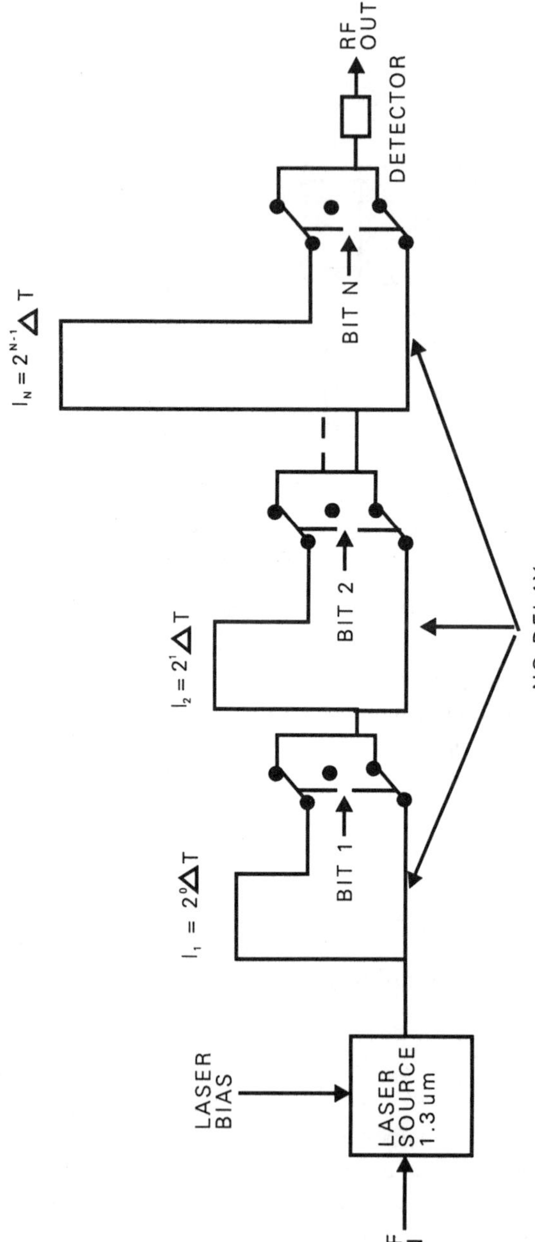

Figure 6.10 Schematic of a binary cascade with *N*-bits resolution. (© 1992 SPIE.)

the number of delay lines including the numbers of laser and photodetector diodes needed to run the time-shifter network. Ng et al. [13] fabricated low-loss GaAlAs/GaAs rib waveguides for optical delay lines in the integrated time-shifter network. A typical cross section of the GaAs/AlGaAs electro-optic waveguide is shown in Figure 6.11. The waveguide is of the strip-loaded variety since etch depths, which give adequate confinment without two-moded guiding, are found to be a little less than the upper cladding thickness. A detailed description of this device is available in [14]. Y-branch cascade splitters are developed to split the optical signal. Waveguide-integrated detectors are developed to perform the function of a switch in the network (100 μm × 15 μm). Table 6.1 shows the component performance of the GaAlAs/GaAs optical time-shifter network.

In Table 6.2, Goutzoulis et al. [12] compared the fiber segment and switch complexities (M_f = number of fibers and M_s = number of switches) as well as the dynamic range loss DR_{loss} of the SRODEL and BIFODEL architectures when used to implement a delay line with delay resolutions of 8 bits, 10 bits, and 12 bits. The BIFODEL architecture significantly outperforms the SRODEL architecture in both M_f and M_s. The dynamic range loss in both architectures is equal and, therefore, both suffer the same DR_{loss}. The data of Table 6.2 indicate that the BIFODEL architecture is the optimum choice for implementation of delay-compressive delay lines.

Using the delay-compressive technique alone, K different BIFODELs are required to steer a PAA of K elements. However, this still corresponds to a significant amount of hardware, and a more efficient architecture must be introduced to further reduce the hardware requirements. For a K-element partitioned PAA, the E sets of N elements can be related by

$$K = N \times E \tag{6.11a}$$

In the case of a one-dimensional PAA, the delay required by the ith element of the jth set is equal to the delay of the ith element of the first set plus a bias delay. The bias delay depends only on j and not on i and, thus, is common to all the elements of a given set. The

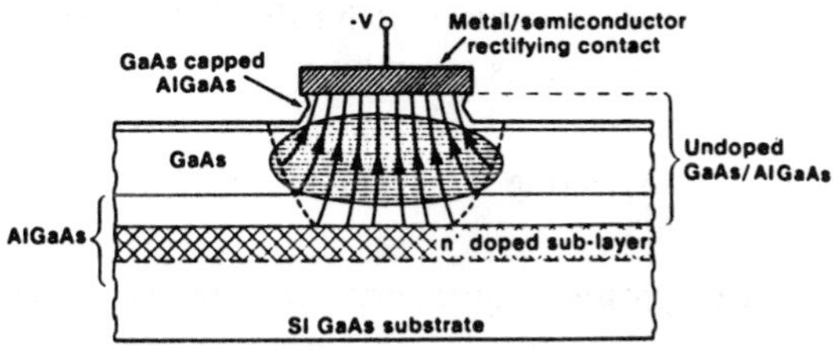

Figure 6.11 Schematic cross section of a typical GaAs/GaA1As electro-optic waveguide.

Table 6.1
Component Performance of a 2-Bit Time Shifter on a GaAs Substrate

Component	Performance
GaAlAs/GaAs rib Waveguide	Refractive index = 3.44 at 1.3-μm wavelength The time delay = 115 psec/cm Losses = 0.3 dB/cm for straight rib = 1 dB/cm for S-bends (curvature larger than 2 mm)
1 × 4 optical splitter	Losses ≤ 1.5 dB/stage and splitting uniformities of better than 0.5 dB
Waveguide-integrated MSM* switch	Switch response = 0.58 mA/mW RF on/off ratio = 40 dB at 2 GHz and 10 GHz for 100-μm-long device

Note: MSM stands for metal-semiconductor-metal.

Table 6.2
Comparison of SRODEL and BIFODEL for *R* = 8 Bits, 10 Bits, and 12 Bits

R, Bits	Parameters	SRODEL	BIFODEL
8		32	8
10	M_f	64	10
12		128	12
8		30	8
10	M_s	62	10
12		126	12
8		9	9
10	DR_{loss} (dB)	10	10
12		11	11

Note: M_f is the number of fibers, M_s is the number of switches, and DR_{loss} is the dynamic range loss.

total number of different types of BIFODELs is $M + E$, which represents a significant reduction in hardware complexity in terms of both BIFODEL type and quantity. Presently, the overall hardware complexity (N_c) is given by

$$N_c = (\log M) \times 2[(K)^{1/2} - 1] \tag{6.11b}$$

The straightforward noncompressed implementation of the hardware complexity, N_u, is given by

$$N_u = K \times M \tag{6.12}$$

Equation (6.11b) contains a logarithmic term and, therefore, provides a reduction in hardware complexity as compared to (6.12). Table 6.3 shows some examples of reductions in the hardware complexity of a delay line (N_c versus N_u). The numbers of hardwares required in the case of BIFODEL (N_c) are shown in Table 6.3 for various array configurations.

6.5 DESIGN OF A SINGLE BIFODEL SEGMENT

Goutzoulis et al. [12] described a block diagram of a single electronically switched BIFODEL segment. Figure 6.12 shows a power-balanced configuration in which the 0-state path between the two back-to-back connected 1 × 2 switches includes an attenuator and the 1-state path between switches contains electronic gain elements to compensate for the losses suffered in optical conversion and transmission. The gain elements in the 1-state path contain an APD detector and an amplifier (AMP-1). The balancing of the two interswitch paths is achieved by combining a small fixed attenuator (2 to 3 dB) in the 0-state path and the APD gain control in the 1-state path. The second amplifier (AMP-2), at the output of the switch combination, compensates for the insertion losses of the two switches, which provides a constant power level to the next segment of the line. Table 6.4 provides information on various components and their specifications that were used in the design.

Table 6.3
Savings of Hardware

Arrays	BIFODEL Case (N_c)	Savings (N_c versus N_u)
12 × 12	24	83.3%
32 × 32	64	96.9%
70 × 70	140	97.1%

Note: N_c is the number of hardware required in the case of a BIFODEL arrangement to steer the antenna array.

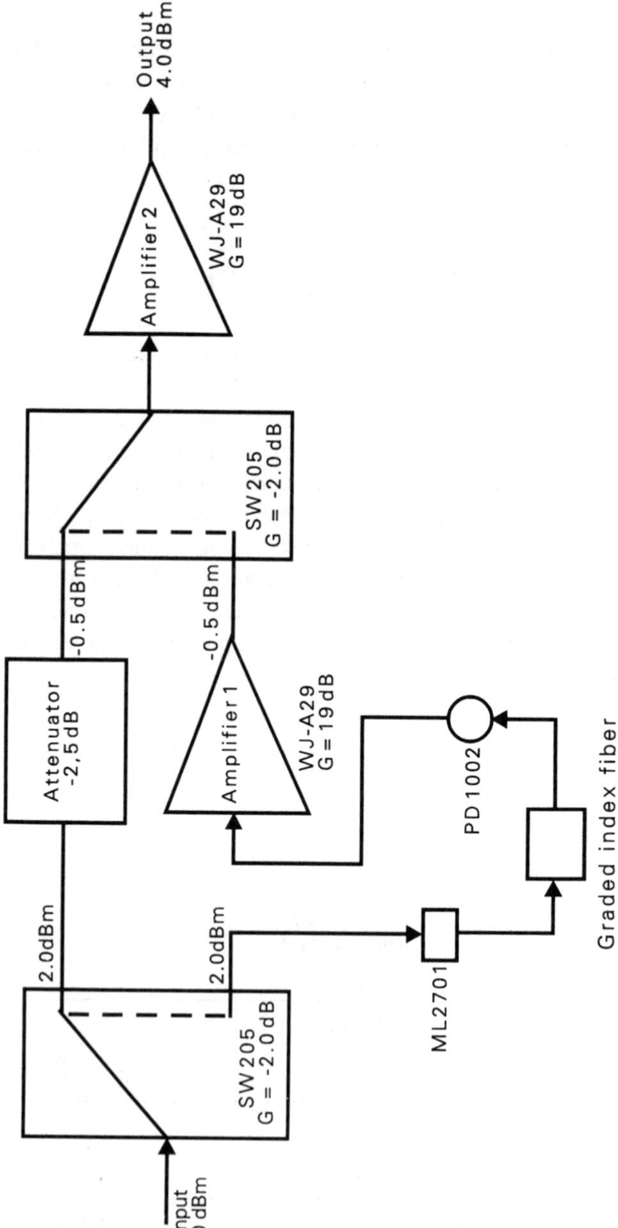

Figure 6.12 A block diagram of a single electronically switched BIFODEL segment. (© 1989 SPIE.)

Table 6.4
Components and Specifications of a Single Electronically Switched BIFODEL Segment

Components	Type and Manufacturer	Specifications
(1) 1 × 2 switch	SW-205 (GaAs MESFET) Adam–Russell	3-dB beamwidth = 5 to 4000 MHz; $T_{sw} < 25$ ns; Maximum input power P_{in} = 25 dBm; Crosstalk at an input level of 25 dBm = −55 dB (single-switch and −65 dB (back-to-back); Third-order intermodulation products IMP at an input level of 10 dBm = −57 dBc.
(2) Laser diodes LDs	ML-2701 LD, Mitsubishi	Frequency = 890 nm; CW Power = 7 mW to 10 mW; Frequency response = flat to within + −1 dB over 0.5 to 1.1 GHz, 3-dB bandwidth extends to 1.3 GHz; R_{IN} = −140 dB/Hz to −146 dB/Hz over the 0.5-GHz to 1-GHz band; IMP = −45 dBc to −55 dBc.
(3) Detector	PD 1002 APD, Mitsubishi	3-dB frequency cut-off = 1.1 GHz to 1.35 GHz for G = 100—well below (by 8 dB) the R_{IN} level of the ML2701 LD; Diameter = 200 μm.
(4) Fiber	GI fiber, Corning	Multimode core diameter = 50-μm/125-μm cladding; Bandwidth = 1584 MHz/km at 850 nm; 3-dB cut-off frequency = 1400 MHz for a length of 1-km multimode fiber; Loss = 2.8 dB/km; The fibers are wound on a Plexiglas™ spool of diameter = 16 cm (to prevent microbend losses); Coupling losses less than 2.5 dB.
(5) Amplifier AMP-1	WJ-A26, Watkins–Johnson	Typical gain = 19 dB over the 10-MHz to 1600-MHz band; Gain flatness = ±0.4 dB; Noise figure = 5 dB.
(6) Amplifier AMP-2	WJ-A29, Watkins–Johnson	Typical gain = 6.5 dB; Gain flatness = ±0.2 dB; Noise figure = 9 dB.

6.6 DELAY-COMPRESSIVE FIBER OPTICAL DELAY LINE

This section describes the element and delay-compressive one-dimensional harware-compressive TTD architecture, two-dimensional hardware-compressive TTD architecture, the two-dimensional array antenna partition technique for independent multibeam formation, the use of directional coupler, and the 1 × 2 (single pole–double throw) switch for T/R operation.

6.6.1 Element- and Delay-Compression for a One-Dimensional Antenna Array

Figure 6.13 shows a block diagram of a BIFODEL-based compressive delay line architecture in the transmit mode. In the figure, the input radar signal modulates the intensity of N_{LD} each at a different optical wavelength (λ). The optical signal from element 1 propagates through the nondelayed path at wavelength λ_1, whereas the remaining $(N - 1)$ signal propagates (at wavelengths λ_i, $i = 2, 3, \ldots, N$) through $(N - 1)$ BIFODEL

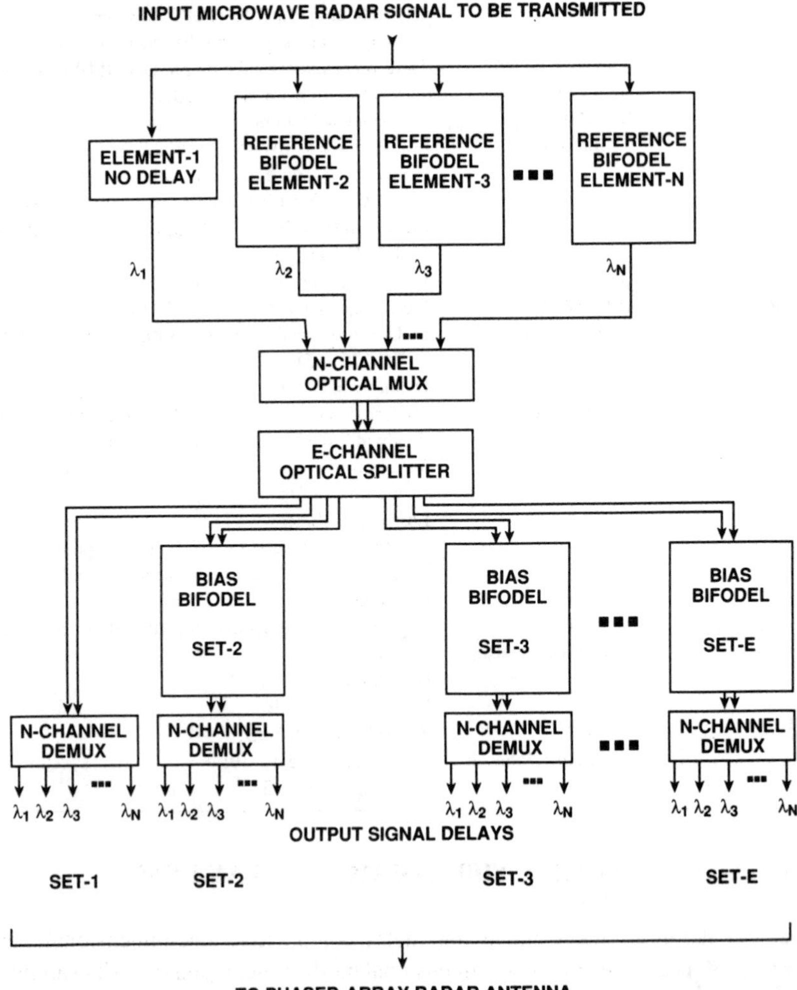

Figure 6.13 A block diagram of a BIFODEL-based compressive delay line structure in transmit mode. (© 1992 SPIE.)

elements 2, . . . , N. All these signals (from nondelayed path and BIFODEL elements 1, . . . , N) are multiplexed at the N-channel optical *multiplexer* (MUX) and then divided into an E-channel optical splitter. In set 1 the signal from the splitter is demultiplexed through an N-channel *demultiplexer* (DEMUX) and the outputs are denoted as $\lambda_1, \ldots, \lambda_N$. The remaining $(N - 1)$ signals from the splitter drive $(N - 1)$ different bias BIFODEL and output signals are demultiplexed at the N-channel demultipliers (sets 2, . . . , E). The output of set 1 demultiplexed without delay and the remaining $(N - 1)$ (sets 2, . . . , E) outputs are delayed via the bias BIFODELs and corresponds to a different PAA set.

Figure 6.14 shows the architecture of the receive mode, which is the reverse of the transmit mode. The output of each antenna element drives an LD of a different wavelength. The outputs of these LDs are multiplexed at the N-channel optical MUX and pass through bias BIFODEL (except set E). After the N-channel MUX, set E signals go directly to the E-channel optical combiner without bias delay. All signals are combined at the E-channel optical combiner and then demultiplexed at the N-channel DEMUX. Each of the DEMUX output signals is fed to reference BIFODELs, and the outputs of the BIFODELs are combined at an RF/microwave combiner. The output of the combiner is the vector sum of the combined RF/microwave signal.

6.6.2 Two-Dimensional Hardware-Compressive TTD Steering Architecture

Recently, Goutzoulis et al. [6, 15] reported the development of a two-dimensional hardware-compressive TTD steering architecture. Figure 6.15 shows the architecture in transmit mode, and the RF/microwave signals are fed at the input. The RF/microwave signals are modulated with the optical signal at LDs and fed to the X-control of the BIFODELs. The outputs of the X-control are multiplexed at the MUX couplers. The signals from the MUX couplers are fed to the Y-control BIFODELs and then demultiplexed. Each signal has a delay of T_{ij} expressed as

$$T_{ij} = i\ dx \sin(\Theta_x)/2c + j\ dy \sin(\Theta_y)/2c$$
$$(i = 1, 2, \ldots, K_x \text{ and } j = 1, 2, \ldots, K_y) \tag{6.13}$$

where dx is the element spacing along x, dy is the element spacing along y, Θ_x is the angular component along x ($\Theta_x = f(x)$), Θ_y is the angular component along y ($\Theta_y = f(y)$), c is the velocity of light, and i and j factors depend on the location of the X- and Y-control BIFODELs.

Figure 6.16 shows the two-dimensional hardware-compressive TTD steering architecture in the receive mode. The signals received from antenna elements are multiplexed and fed to the Y-control of the BIFODEL. The output signals of the Y-control of the BIFODEL are demultiplexed and then fed to the X-control BIFODELs. Finally, these signals are fed to the corresponding PDs and the RF/microwave signals are received at the output. This process is the reverse of the transmit process.

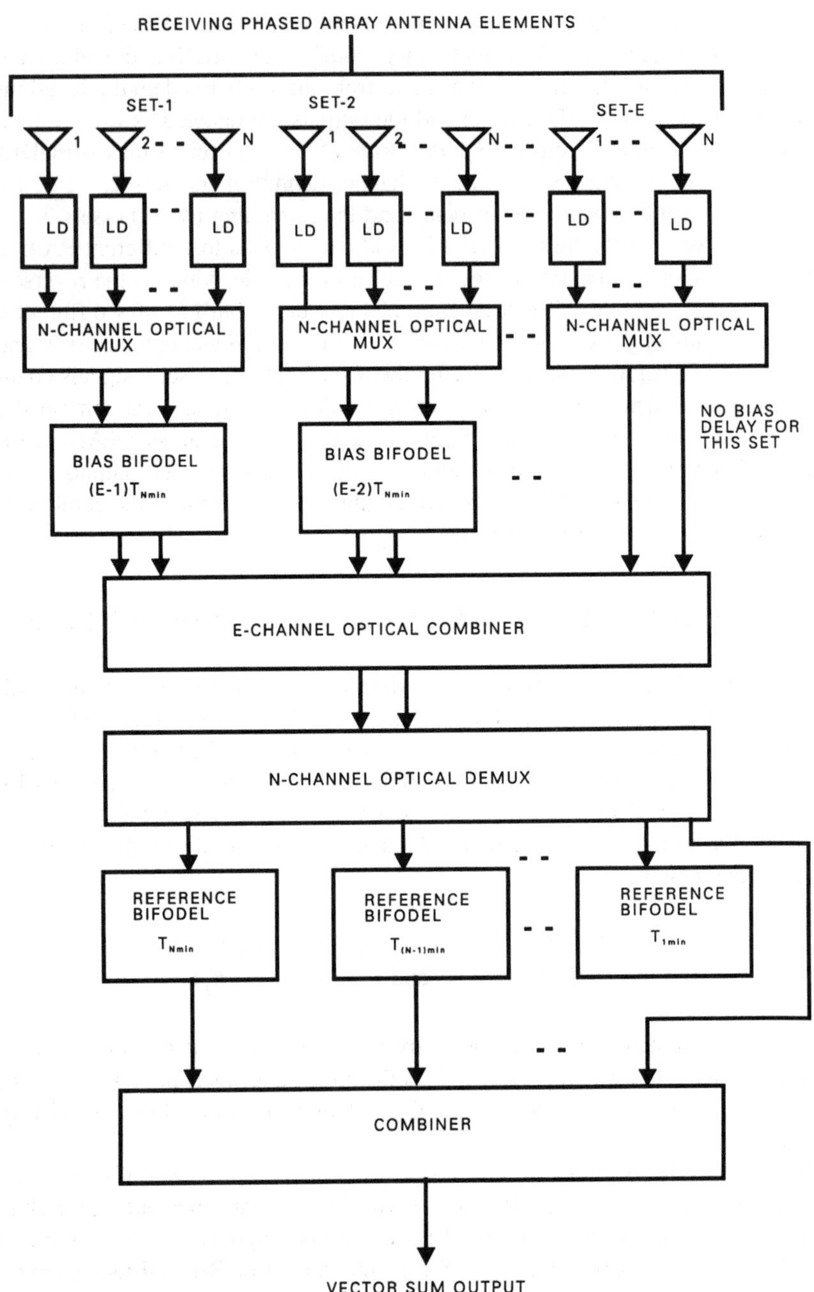

Figure 6.14 A block diagram of a BIFODEL-based compressive delay line structure in receive mode. (© 1990 OSA.)

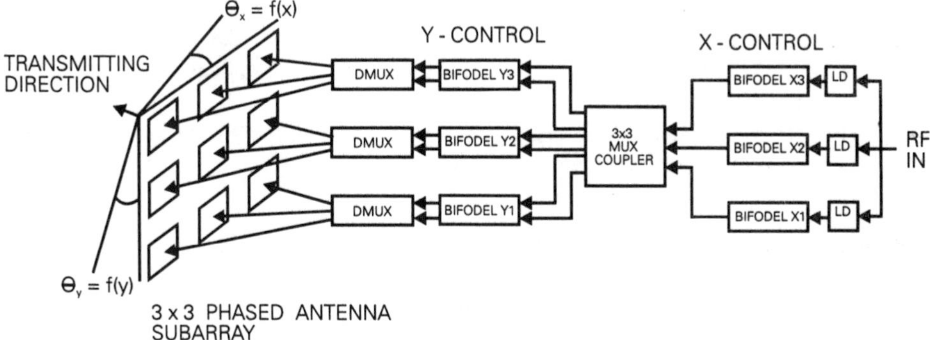

Figure 6.15 Two-dimensional hardware-compressive TTD steering architecture in transmit mode. (© 1992 SPIE.)

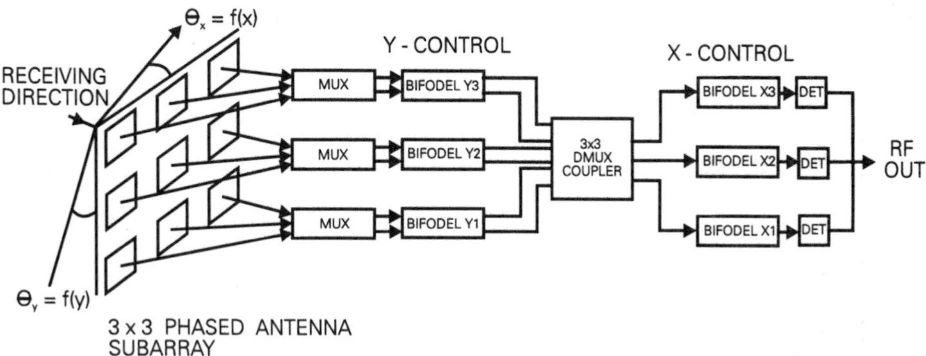

Figure 6.16 Two-dimensional hardware-compressive TTD steering architecture in receive mode. (© 1992 SPIE.)

Implementation of the T/R Operation

The minor hardware additions are needed to implement the T/R operation in the two-dimensional hardware-compressive TTD steering architecture. There are two methods for this implementation, namely, (1) the *directional couplers* (DCs) method, and (2) the 1 × 2 switch method. Figure 6.17(a) shows the architecture of the DCs method and Figure 6.17(b) shows a block diagram of the 1 × 2 switch method. In both methods, each T/R model can (1) detect light and amplify the resulting electrical signal, and (2) produce a modulated signal via its own LD at the proper wavelength.

In Figure 6.17(a), a DC is used at each T/R module and at the input of the *X*-control (K_x) BIFODELs. In the transmit mode, the optical signal propagates toward the T/Rs from the input and is directed to the PD of each T/R as in the receive mode, the optical signal propagates backward from each T/R LD to the PD at the input as shown by the direction of in the figure. In the case of the 1 × 2 switch method, a 1 × 2 switch module is used at each T/R section as well as at the input of the *X*-control BIFODELs. The switch setting changes for transmit and receive operations; the directions of the optical signals are shown by the arrows in the figure.

The DC is a passive device and is less expensive than the switch and is therefore more preferable than the 1 × 2 switch method.

Advantage of the Two-Dimensional Hardware-Compressive TTD Steering Architecture

Mathematically, the BIFODEL complexity (C_b) is given by

$$C_b = K_x + K_y \tag{6.14}$$

From the above equation, we can provide some practical examples of hardware savings as shown in Table 6.4.

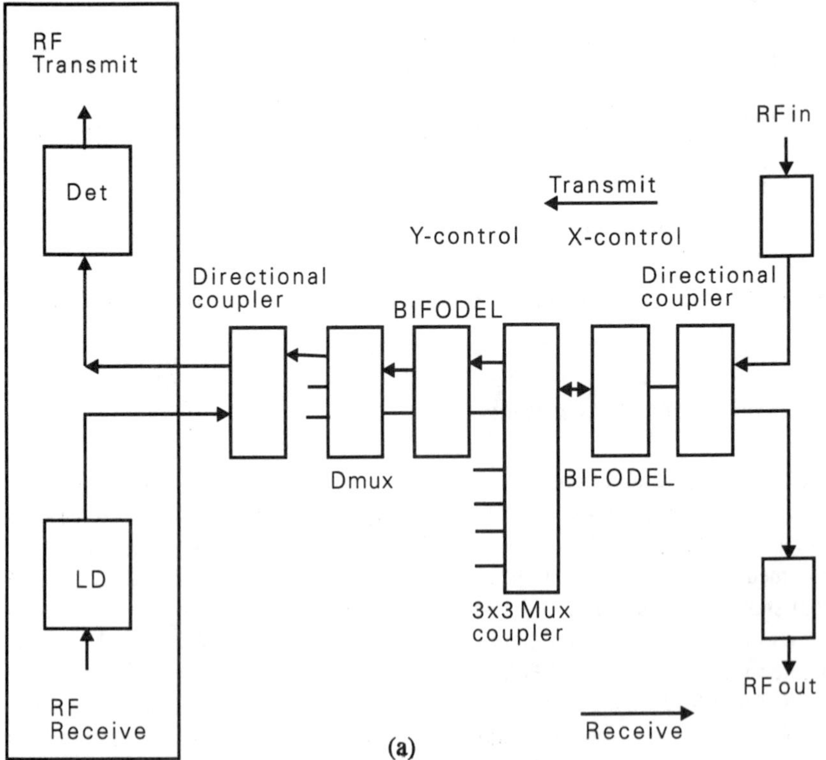

Figure 6.17 (a) Directional coupler and (b) 1 × 2 switch approaches for T/R operation. (© 1992 SPIE.)

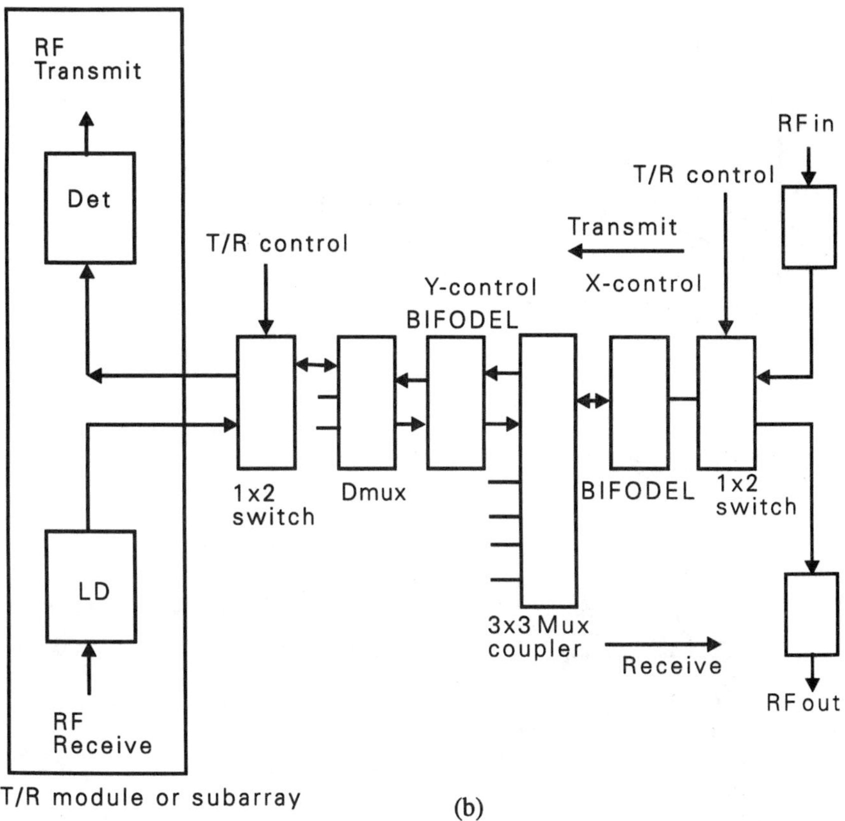

Figure 6.17 (continued)

The Two-Dimensional Array Antenna Partition Technique for Independent Multibeam Formation

In this discussion, multiple simultaneous beams at different directions require the simultaneous use of delay combinations, which is not possible without increasing the hardware (minimum twice) with BIFODELs that employ DC/2-state switches. A two-dimensional partition technique was reported by Goutzoulis et al. [15, 16] that provides the BIFODEL complexity (C_m),

$$C_m = K_x + mK_y \tag{6.15}$$

where m is the independent beams. Figure 6.18 shows a two-dimensional array antenna

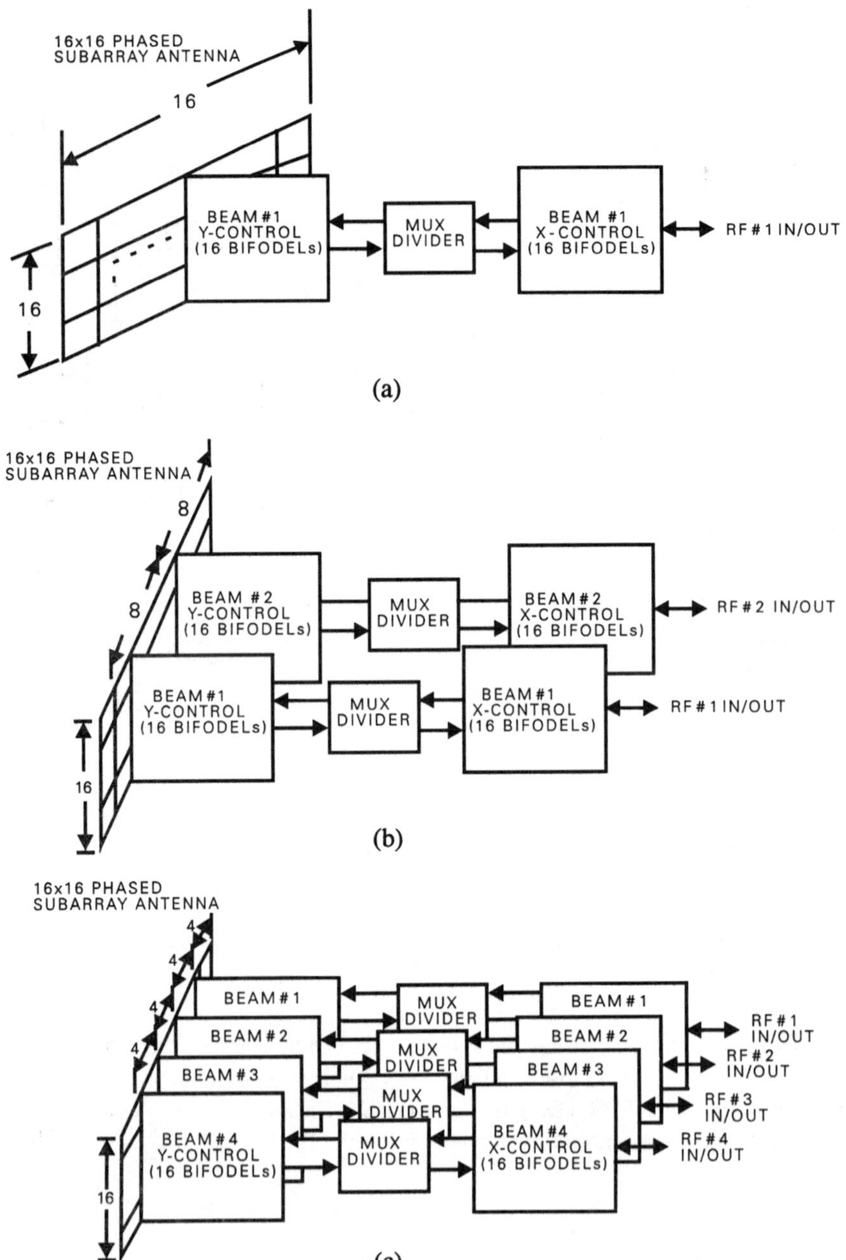

Figure 6.18 Two-dimensional array antenna partition technique for independent (a) single-beam, (b) dual-beam, and (c) quadruple-beam operation. (© 1992 SPIE.)

partition technique for a 16×16 array that generates 1, 2, and 4 independent beams in both frequency and direction. In this technique, the main idea is to use the same number of *X*-control BIFODELs and double the number of *Y*-control BIFODELs every time we require doubling the independent beams as shown in Table 6.5.

System Control

Overall system control is simple because all BIFODELs for *X*- and *Y*-controls require the identical binary program. The main reason for this is that if we want to address a 6-bit system, then we generate two 6-bit digital control words each of which is applied in parallel to the BIFODELs within the *X*- and *Y*-groups. The 6-bit words are the binary representations of the desired angles Θ_x and Θ_y and are independent of the number or location of the antenna array elements. The finite wavelength spacing limits the size of the λ-MUX TTD system.

Experimental Results

Goutzoulis et al. [16] constructed a 16-element phased array antenna driven by the WDM TTD system and reported squint-free antenna array patterns at −45°, 0°, and +45°. Twelve superimposed antenna patterns are shown in Figure 6.19 for frequencies of 600 MHz, 900 MHz, 1200 MHz, and 1500 MHz. They also performed the experiments to verify the angular resolution of the TTD system. The even-numbered elements of the 16-element antenna array with a fixed input frequency of 700 MHz are used. In the experiments, a total of 63 patterns are recorded, one for each binary combination of the six-bit switches.

Table 6.5
Number of Beams and BIFODELS *X*- and *Y*-Controls

Number of Beams	*X*-Control BIFODELs	*Y*-Control BIFODELs
2	2 sets of 8 = 16	2 sets of 16 = 32
4	4 sets of 16 = 64	4 sets of 16 = 64

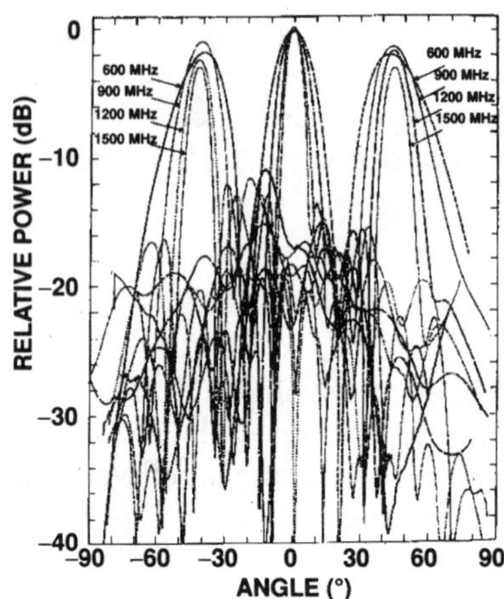

Figure 6.19 Squint-free radiation patterns of the antenna for steering at −43°, 0°, and +45°. [16].

6.7 CONCLUSIONS

This chapter provides a review of binary optical fiber delay lines. It is shown that the reduction in hardware using BIFODEL is more than 83 percent for a small array of 12 × 12. The hardware reduction for a large-sized array such as 70 × 70 is more than 97 percent. Goutzoulis et al. [16] reported the development of three different generations of electronic delay lines (*DiBi*) (these delay lines are not discussed in the chapter) and compared the characteristics and costs with the BIFODELs. The time-delay performance of the BIFODELs is found to be similar to that of the *DiBi*'s. However, the BIFODEL can provide longer delays (hundreds of microseconds) and are transparent to microwaves, whereas the *DiBi* is limited to a few nanoseconds and few gigahertz bandwidth. Squint-free antenna-array patterns at −43°, 0°, and +45° are described, and it is possible to use this type of system for a large array antenna.

References

[1] Ng, W., A. Walston, G. L. Tangonan, I. Newberg, and J. J. Lee, "Wideband Fiber-Optic Delay Network for Phased Array Antenna Steering," *Electron. Lett.*, Vol. 25, No. 21, October 1989, pp. 1456–1457.

[2] Ng, W., A. Walston, G. L. Tangonan, J. J. Lee, I. L. Newberg, and N. Bernstein, "The First Demonstration of an Optically Steered Microwave Phased Array Antenna Using True-Time-Delay," *J. Lightwave Tech.*, Vol. 9, No. 9, September 1991, pp. 1124–1131.

[3] Ng, W., A. Walston, G. Tangonan, J. J. Lee, and I. Newberg, "Optical Steering of Dual Band Microwave Phased Array Antenna Using Semiconductor Laser Switching," *Electron. Lett.*, Vol. 26, No. 12, June 1990, pp. 791–793.

[4] Kondo, M., K. Komatsu, Y. Ohta, S. Suzuki, K. Nagashima, and H. Goto, "High-Speed Optical Time Switch with Integrated Optical 1 × 4 Switches and Single-Polarization Fiber Delay Lines," *Tech. Digest,*

Fourth Int. Conf. on Integrated Optics and Fiber Optic Communication, Tokyo, Japan, Paper 29D3–7, 1983.

[5] Soref, R. A., ''Programmable Time-Delay Device,'' *Appl. Opt.,* Vol. 23, No. 21, November 1984, pp. 3736–3737.

[6] Goutzoulis, A., and D. Davies, ''Hardware Compressive 2-D Fiber Optic Delay Line Architecture for Time Steering of Phased Array Antennas,'' *Appl. Opt.,* Vol. 29, 1990, pp. 5353–5359.

[7] Levine, A. M., ''Fiber-Optic Phased Array Antenna System for RF Transmission,'' U.S. Patent No. 4028702, 1977.

[8] Levine, A. M., ''Use of Fiber Optic Frequency and Phase Determining Elements in Radar,'' *Proc. 33rd Symp. for Frequency Control,*'' 1979, pp. 436–443.

[9] Hemmi, C. O., and C. E. Takle, ''Optically Controlled Phased Array Beamforming Using Time Delay,'' *Proc. SPIE,* Vol. 1703, 1992, pp. 545–550.

[10] Taylor, H. F., ''Optical Fiber Devices for Signal Processing,'' *Proc. SPIE,* Vol. 209, 1980, pp. 159–166.

[11] Lagerstroem, B., P. Svensson, A. Djupsjoebacka, L. Thylen, and B. Stoltz, ''Integrated-Optic Delay Line Signal Processor,'' *Proc. OFC/IOOC,* 1987, Paper WK2, p. 176.

[12] Goutzoulis, A. P., D. K. Davies, and J. M. Zomp, ''Prototype Binary Fiber Optic Delay Line,'' *Opt. Engrg.,* Vol. 28, No. 11, November 1989, pp. 1193–1202.

[13] Ng., W., D. Yap, A. Narayanan, R. Hayes, and A. Walson, ''GaAs Optical Time-Shifter Network for Steering a Dual-Band Microwave Phased Array Antennas,'' *Proc. SPIE,* Vol. 1703, 1992, pp. 379–383.

[14] Boyd, J. T., Ed., *Integrated Optics: Devices and Applications,* Piscataway, New York: IEEE Press, 1990.

[15] Goutzoulis, A. P., and D. K. Davies, ''All-Optical Hardware-Compressive Wavelength-Multiplexed Fiber Optic Architecture for True Time Delay Steering of 2-D Phased Array Antennas,'' *Proc. SPIE,* Vol. 1703, 1992, pp. 604–614.

[16] Goutzoulis, A. P., D. K. Davies, J. Zomp, P. Hrycak, and A. Johnson, ''A Hardware-Compressive Fiber-Optic True Time Delay Steering System for Phased Array Antennas,'' *Microwave J.,* Vol. 37, No. 9, September 1994, pp. 126–140.

Chapter 7
Optical Beam Steering of an Antenna Array Using Two Lasers

7.1 INTRODUCTION

Various techniques were investigated in Chapters 4, 5, and 6 that require one optical frequency laser source. In this chapter, we describe a two-laser source model to control an array antenna using Fourier transfer and optical/electrical conversion. A two-laser model using two laser diodes (sources) whose frequency difference is set to a desired RF/ microwave frequency is proposed. Relations among antenna radiation characteristics such as gain, beamwidth, the carrier-to-noise ratio of the RF/microwave signal, and reduction of the antenna sidelobe are discussed for an optically controlled array antenna.

7.2 CONCEPT OF AN OPTICALLY CONTROLLED ARRAY

The two-laser model concept shown in Figure 7.1 is described by Konishi et al. [1–4] and Pan et al. [5–7]. Two laser diodes LD_1 and LD_2 are shown as two optical frequency sources. The light beam emitted from LD_1 passes through an optical lens and a pinhole image mask. The beam from the pinhole mask is incident on a *Fourier transformation* (FT) lens. The Fourier-transformed light beam (P_1) is combined with the light beam of the plane wave (P_2) from LD_2 and sampled especially with an optical fiber array that is set on the focal plane of the FT lens. A photodiode detector is placed in the end of each fiber to generate a frequency difference between P_1 and P_2. The difference between P_1 and P_2 is the RF/microwave signal at the output of each photodiode detector. The RF/microwave signal from photodiode detectors is fed to each antenna element via an amplifier. Phase and amplitude information of laser lightwave generated and controlled by the pinhole image mask and FT lens is then transferred to the antenna array. The position and shape of the lightwave passing through the pinhole image mask correspond to the beam direction and the far-field pattern of the RF/microwave radiated from the array.

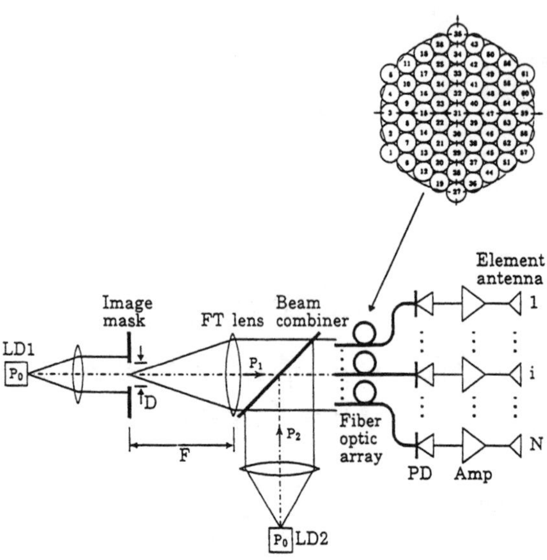

Figure 7.1 Two-laser model diagram [1].

The antenna radiation pattern becomes a fan beam for an antenna array when the mixed beam is sampled by the fiber array. The gain (G) and beamwidth Θ are given by [1]

$$G = \frac{1}{m} \cdot \frac{2F}{D} \tag{7.1}$$

$$\Theta = 2\sin^{-1}\left(m \cdot \frac{D}{2F}\right) \tag{7.2}$$

$$m = \frac{d_f/\lambda_o}{d_a/\lambda_m} \tag{7.3}$$

where D is the aperture diameter of a pinhole image mask, F is a focal length of a FT lens, d_a is the element spacing of the fiber array, d_m is the element spacing of antenna elements, λ_o is an optical wavelength, and λ_m is a RF/microwave frequency wavelength.

7.3 PINHOLE IMAGE MASK

The pinhole image mask is one of the key components in the Fourier optic processor. An optical signal passes through the output of the pinhole mask if the input and output are aligned. In the case of misalignment, the input signal is blocked and works like the off position of the switch.

There are several techniques for constructing the pinhole image mask. These are: mechanical pinhole [8], liquid crystal light valve [9], magnetic optical light modulator [10], holograph, vector multiplier, acousto-optic gate, and vertical cavity surface-emitting array. However, none of these devices can achieve better than a 25-dB on-off ratio except the mechanical pinhole image mask. The switching times of these devices are not less than 20 μs. The control speed of the mechanical image mask is very slow (about 1 μs). The limited on-off ratio of the image tends to broaden the mainlobe and heighten the sidelobe levels of the RF/microwave far-field radiation pattern. Anderson et al. [8] investigated to find the lower limit of a contrast ratio adequate for far-field radiation patterns. The criteria for the evaluation were sidelobe levels and mainlobe beamwidths. It was determined that a minimum contrast ratio of 30 dB was required for spot beam and a 25-dB level for fan beams.

7.4 DESIGN OF A FIBER OPTIC SHUTTER SWITCH

An optical fiber shutter switch was reported by Pan et al. [7] that can provide a 50-dB on-off ratio with switching time less than 1 ns. Figure 7.2 shows a diagram of a fiber optic shutter switch. The switch is constructed with two GRIN lenses, two electro-optic (EO) wedges ($LiNbO_3$) with opposite optical axes, and a controller. The length and the width of the EO wedge are b and a, respectively; and the diameter of the laser beam is D, as shown in the figure. When a voltage v is applied to the EO, $n \neq 0$, and therefore, the passing light through the EO is tilted at an angle. The relationship of the angle is calculated by simple optics and is given by

$$\theta = \mathrm{tg}^{-1}\left(\frac{a}{b}\right) \tag{7.4}$$

$$\theta_2 = \sin^{-1}\left[\frac{n + \Delta n}{n - \Delta_n} \sin\,(90° - \theta)\right] \tag{7.5}$$

$$\theta_3 = \theta_2 - \theta_1 \tag{7.6}$$

$$\theta_4 = \sin^{-1}\,[(n - \Delta n)\,\sin(\theta_3)] \tag{7.7}$$

$$d = e\tan\,(e_3) \tag{7.8}$$

$$e = \frac{(a + D)b}{2a} \tag{7.9}$$

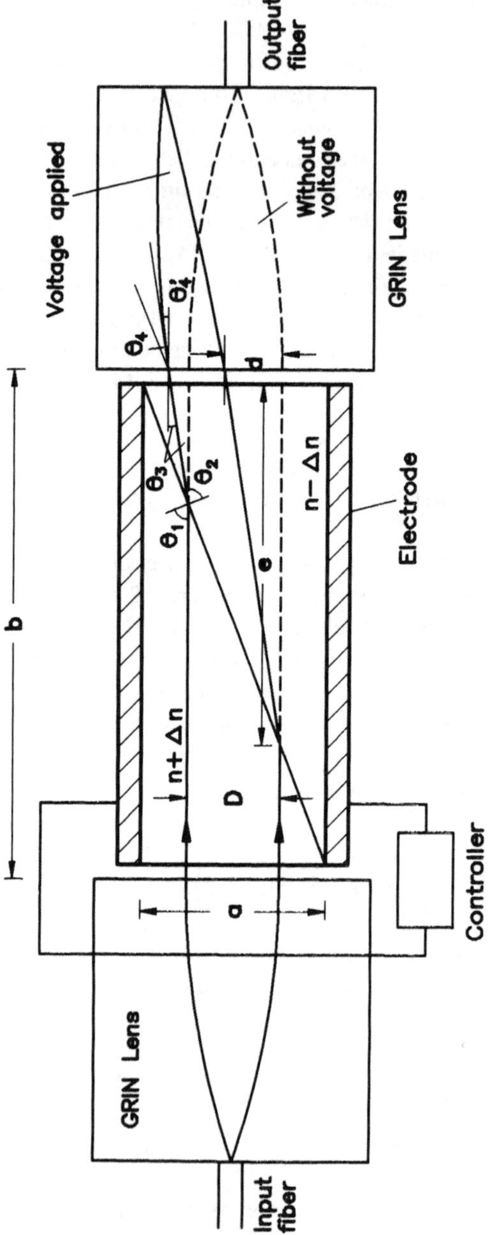

Figure 7.2 A typical optical fiber SBN shutter switch. (© 1991 SPIE.)

where n is the refractive index of the $LiNbO_3$, $\Delta n = 1/2n^3r\ V/a$, v is the applied voltage to EO, and r is the electro-optic coefficient.

When the voltage v is applied to the EO, the light is tilted and will not pass through the output fiber. However, when no voltage is applied, $V = 0$ and the light will pass through the output fiber. This way, the laser light can be switched "on" and "off" by simply controlling the applied voltage.

Pan et al. [7] considered various materials for the switch that is summarized in Table 7.1. The relationship between insertion loss and the separation of lens GRIN is shown in Figure 7.3. Figure 7.4 shows the plot of the offset distance (d) versus the applied voltage (v) for SBN material with width = 0.5 mm, length = 20 mm, and beam diameter = 0.21

Table 7.1
Materials for Fiber Optic Shutter Switch

Materials	n	r_0(pm/v)	Permittivity($\epsilon\gamma$)
KTP	1.86	35	13
$LiNbO_3$	2.20	30	34
Merocyanine/PA-2	1.56	>1000	1.6 to 2.2
$Sr_x\ Ba_{1-x}\ NbO_3$ (SBN, $x = 0.75$)	2.298	1340	3800

Source: [7].

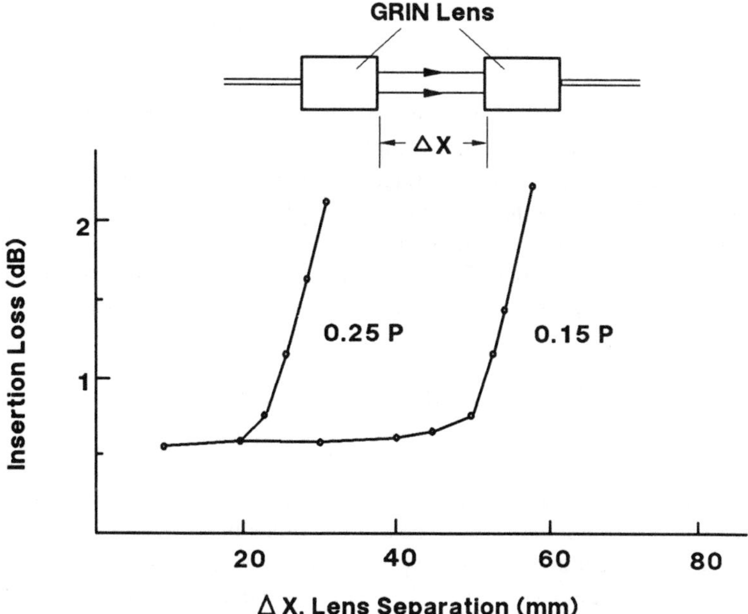

Figure 7.3 Relationship between insertion loss and GRIN lens separation. (© 1991 SPIE.)

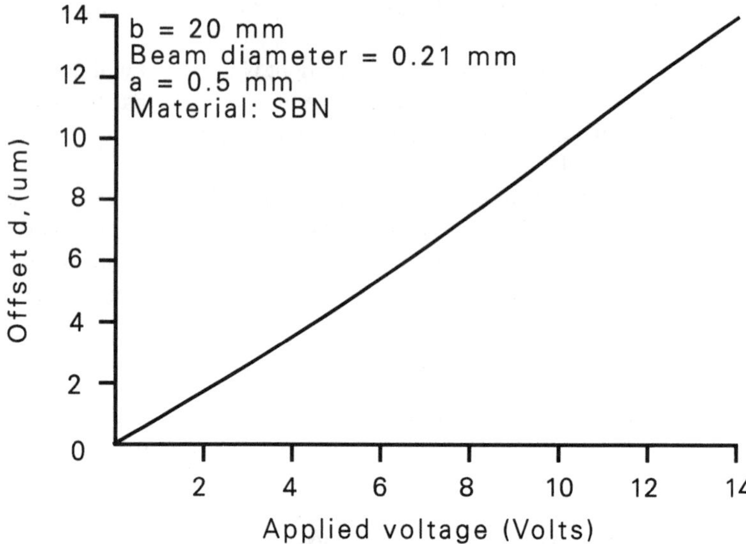

Figure 7.4 Offset of an SBN shutter switch versus applied voltage. (© 1991 SPIE.)

mm. A −50-dB decrease is experimentally obtained for a slant angle of 1.3°; therefore, a 50-dB on-off ratio is achieved for an applied voltage of 14 V to the SBN material. Merocyanine material has the lowest refractive index and relative permittivity in Table 7.1. This material can provide high-speed switching for antenna beam steering and shaping. Fiber optics shutter switch matrices are computer controlled; therefore, it can be used at high-speed on-off ratios, which allows the phased array antenna to perform, for example, beamforming, steering, nulling for fast searching, tracking, and pointing.

7.5 SIGNAL-TO-NOISE RATIO AND LASER OUTPUT

The available carrier-to-noise ratio in 1 Hz (C/N_0) of both microwave signals fed to each element antenna and radiated array antenna is important for communication applications. The image mask aperture is assumed to be a circle with a uniform amplitude distribution. In the case of a two-laser system, the amplitude distribution in the reference beam P_2 is assumed to be Gaussian of exp $(-\beta^2 r^2)$, in which r is the distance from the center of the fiber optic array. The carrier-to-noise ratio in 1 Hz fed to element antenna i, $(C/N_0)_{ele,i}$, is given by [4]

$$\left(\frac{C}{N_0}\right)_{\mathrm{ele},i} = \frac{1}{2}\left(\frac{\pi\beta}{4\lambda_\mathrm{o}}\right)^2 \alpha_1' \, \alpha_2' \, \alpha_3'^2 \, R_\mathrm{L} \, R^2 \, d_\mathrm{c}^4$$

$$\cdot \left(\frac{D}{F}\right)^2 e^{-2\beta^2 r_i^2} \left[\frac{J_1(u_i)}{u_i}\right]^2 \frac{P_\mathrm{o}^2}{P_\mathrm{n}} \tag{7.10}$$

$$u_i = \frac{\pi}{\lambda_\mathrm{o}} \cdot \frac{D}{F} \cdot r_i \tag{7.11}$$

$$\begin{aligned} \alpha_1' &= \alpha_{11} \cdot \alpha_{12} \cdot \alpha_{13} \\ \alpha_2' &= \alpha_{22} \cdot \alpha_{23} \\ \alpha_3' &= \alpha_{31} \cdot \alpha_{32} \end{aligned} \tag{7.12}$$

where R_L is the load resistance for the photodetector diode, R is the sensitivity of photodiode, d_c is the core diameter of the optical fiber, λ_o is the optical wavelength, D is the aperture diameter of the image mask, F is the focal length of the FT lens, r_i is the distance between element i and the center of the fiber optic array, P_o is the respective output power of LD_1 and LD_2, α_{13} and α_{23} are the coupling losses of LD_1 and LD_2, respectively, P_n is the thermal noise of photodiode, α_{21} is the insertion loss of the external frequency modulator, α_{12} is the insertion loss of the external frequency modulator, α_{12} and α_{22} are the respective insertion losses of the beam combiner for the output beam of the FT lens P_1 and for the reference P_2, α_{31} is the coupling loss between the combined beam $P_1 + P_2$ and the fiber optic array, and α_{32} is the connection loss between each optical fiber and photodiode.

Figure 7.5(a) shows plots of the carrier-to-noise ratio of an element versus *D/F* as functions of laser diode output power and edge level of the amplitude distribution in the reference beam (EL). Table 7.2 shows the parameters used for calculation of required laser output power.

The carrier-to-noise ratio in 1 Hz of the microwave signal radiated from the array antenna $(C/N_0)_\mathrm{ant}$ is given by

$$\left(\frac{C}{N_0}\right)_\mathrm{ant} = G \cdot \frac{1}{N} \sum_{i=1}^{N} \left(\frac{C}{N_0}\right)_{\mathrm{ele},i} \tag{7.13}$$

where N is the number of antenna elements and G is the gain of the array antenna. Figure 7.5 (b) shows the carrier-to-noise ratio for the array antenna versus *D/F* as a function of the laser output power and EL. In Figure 7.5(b), $(C/N_0)_\mathrm{ant}$ is saturated at $D/F = 2.5 \times 10^{-3}$, because the beamwidth of the radiation pattern increases and the antenna gain becomes lower as *D/F* increases. However, in the case of Figure 7.5(a), $(C/N_0)_\mathrm{ant}$ is proportional to *D/F*. When the aperture amplitude and phase distributions have Gaussian error ΔA in amplitude and $\Delta \phi$ in phase, plots of $(C/N_0)_\mathrm{ant}$ versus *D/F* are shown in Figure 7.5(c).

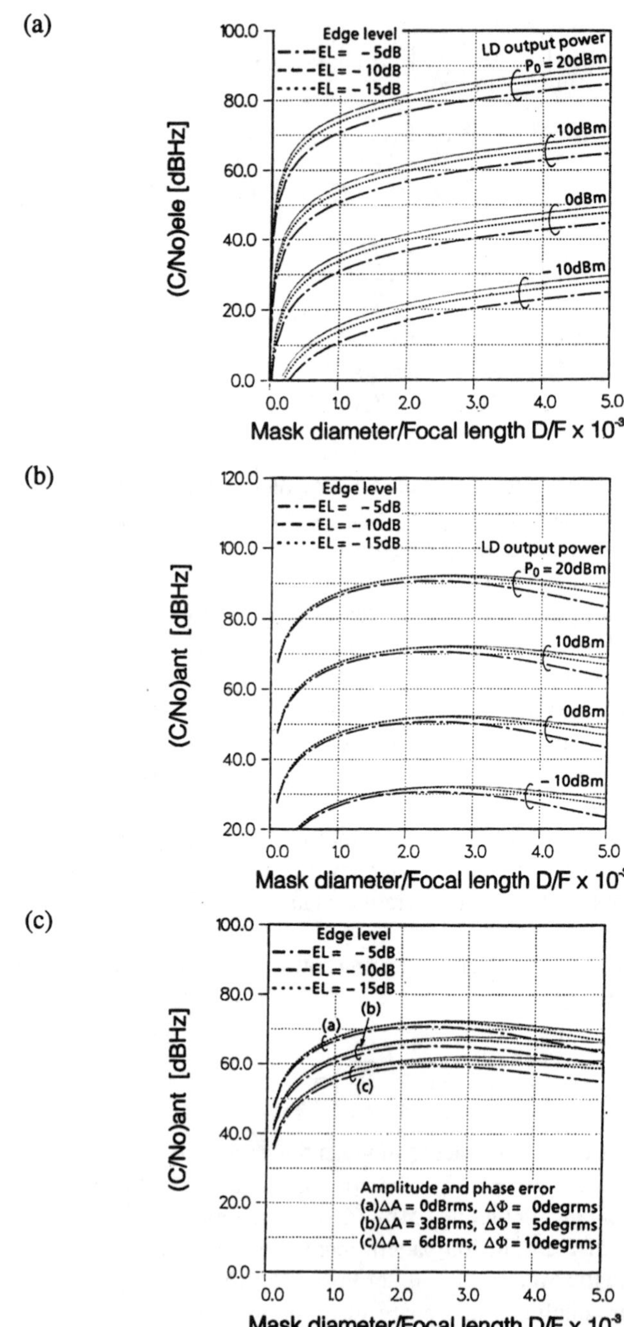

Figure 7.5 (a) Plots of $(C/N_0)_{ele}$ versus *D/F* for different laser diode output power and edge level. (b) Plots of $(C/N_0)_{ant}$ versus *D/F* for different laser diode output power and edge level. (c) Plots of $(C/N_0)_{ant}$ versus *D/F* using a Gaussian error. (© 1992 IEEE.)

Table 7.2
Parameters Used for Analysis of the Antenna Radiation Patterns

Parameters	Value/Description
Amplitude distribution on the mask aperture	Uniform
Optical wavelength (λ_o)	1.3 μm
Mask aperture length/ focal length (*D/F*)	0 to 5×10^{-3}
Amplitude distribution in the reference beam	Uniform and Gaussian
Fiber element spacing (d_f)	125 μm
Antenna element spacing (d_a)	0.5 λ_m
Antenna element type	Isotropic
No. of antenna elements (*N*)	11, 21, and 31

7.6 RADIATION PATTERNS AND REDUCTION IN SIDELOBE LEVEL

The antenna radiation pattern for the antenna system of Figure 7.1 becomes a fan beam, and the calculation is based on the following assumptions.

1. The amplitude distribution in the reference beam P_2 is Gaussian of exp $(-\beta^2 r^2)$.
2. The antenna is linear array.
3. The image mask contains a one-dimensional slit with a uniform amplitude distribution.

The radiation pattern is calculated by [4, 11]

$$F(\theta) = \sum_{i=1}^{N} f(\theta)\, e^{-\beta^2 m^2 (\lambda_o/\lambda_m)^2 x_i^2} \cdot \frac{\sin u'}{u'} \cdot e^{j(2\pi/\lambda_m)x_i \sin\theta} \tag{7.14}$$

$$u' = \frac{\pi}{\lambda_m} \cdot m \cdot \frac{D}{F} \cdot x_i \tag{7.15}$$

$$m = \frac{d_f/\lambda_o}{d_a/\lambda_m} \tag{7.16}$$

where x_i is the position of element antenna *i*, d_f is the fiber element spacing, d_a is the antenna element spacing, and $f(\theta)$ is the element pattern.

Figures 7.6, 7.7, and 7.8 show the calculated radiation patterns for the pinhole diameters of 30 μm, 75 μm, and 100 μm. "Evaluated" and "calculated" radiation patterns are plotted by the calculations on the basis of the measured aperture distributions and the theoretical aperture distributions, respectively. In the case of the pinhole diameter of 30 μm, the evaluated radiation pattern is wider than the theoretical pattern. The shape of the pattern in the central region is asymmetrical for the pinhole diameters of 75 μm and 100 μm. Figure 7.9(a–c) shows radiation patterns when the pinhole is set transversally at

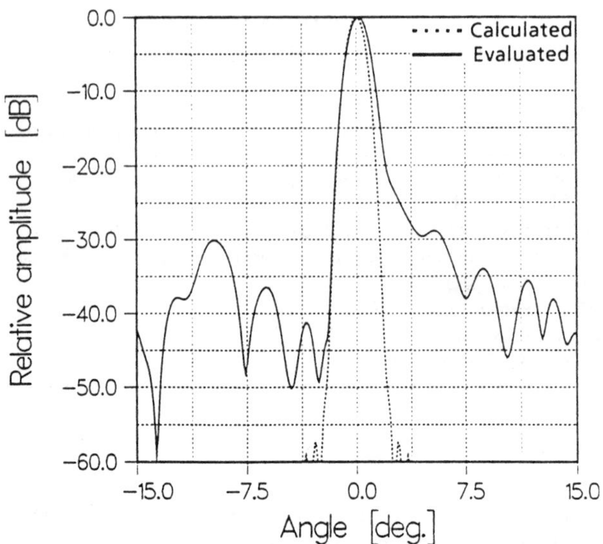

Figure 7.6 Radiation pattern of an array antenna for $D = 30\ \mu m$. (© 1992 SPIE.)

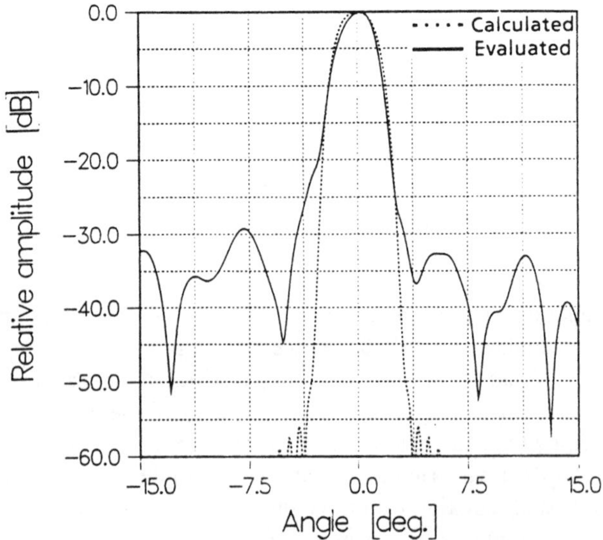

Figure 7.7 Radiation pattern of an array antenna for $D = 75\ \mu m$. (© 1992 SPIE.)

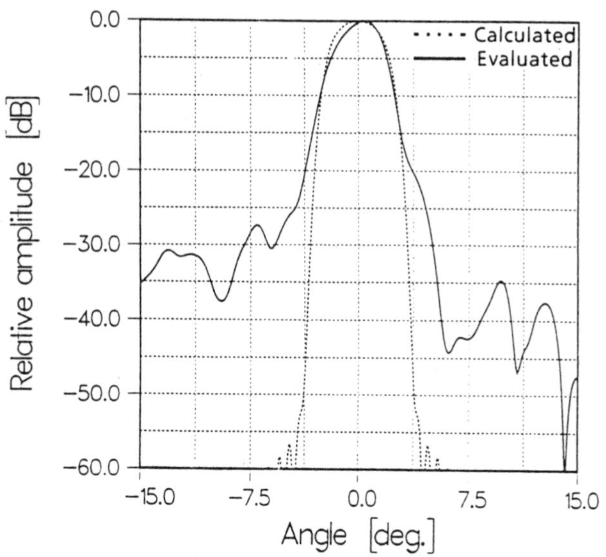

Figure 7.8 Radiation pattern of an array antenna for D = 100 μm. (© 1992 SPIE.)

−50 μm, 0, and +50 μm. The scan angle for the antenna beam is also shown in the figure. The beam is scanned by shifting the pinhole transversally. The scanning angle of ±15° is obtained when sampling interval is 300 μm and the number of antenna elements is 27 in an array.

The beamwidth (Θ) and the gain (G) are obtained by assuming that the antenna contains a large number of isotropic elements, and the P_2 provides uniform amplitude distribution. The combined beam P_1 and P_2 is ideally sampled by the fiber optic array. Θ and G are given by

$$\Theta = 2\sin^{-1}\left(m \cdot \frac{D}{2F}\right) \tag{7.17}$$

$$G = \frac{1}{m} \cdot \frac{2F}{D} \tag{7.18}$$

The relationship between the ratio D/F and the magnification m is calculated on the ideal assumptions to obtain the approximate values of beamwidth and the gain. These assumptions are as follows:

1. The number of elements are finite.
2. The shape of the beam P_2 is in the Gaussian form.
3. There is no sidelobe in the radiation pattern.

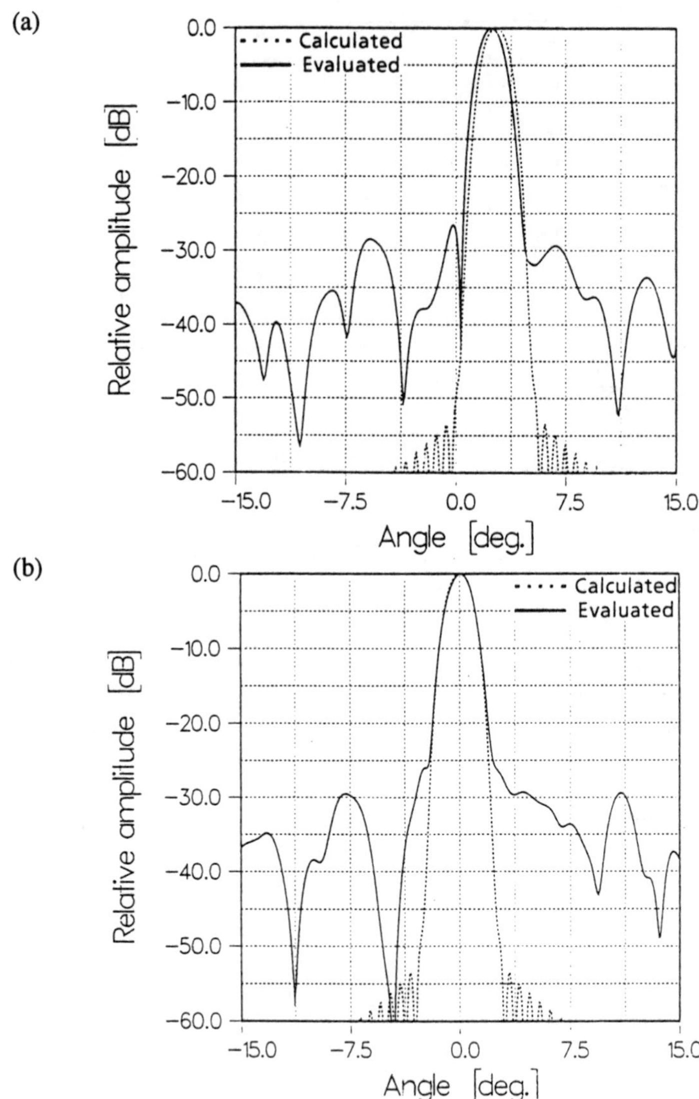

Figure 7.9 Radiation pattern of an array antenna for (a) $X_0 = -50$ μm, (b) $X_0 = 0$ μm [3], and (c) $X_0 = +50$ μm. (© 1992 SPIE.)

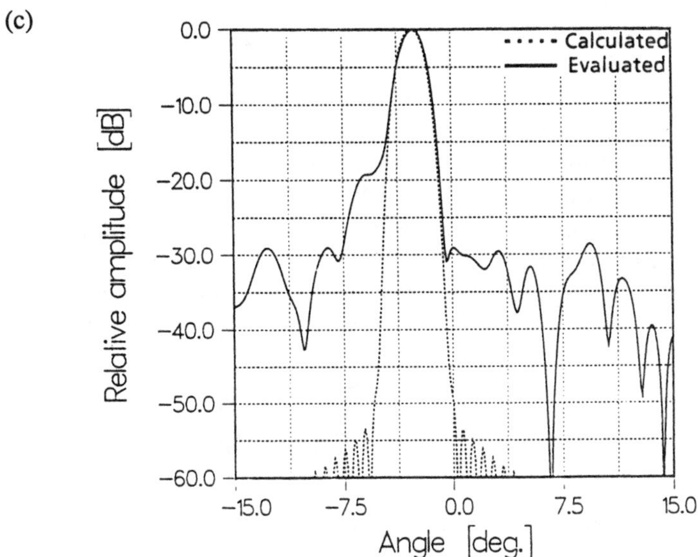

Figure 7.9 (continued)

The relationship between the ratio *D/F* and *m* is given by

$$\frac{D}{F} \leq \frac{2}{m} \tag{7.19}$$

Figure 7.10 shows the beamwidth and gain for the uniform reference beam as functions of elements (*N*) and *D/F*. The "ideal" curves are plotted from (7.17) and (7.18). In Figure

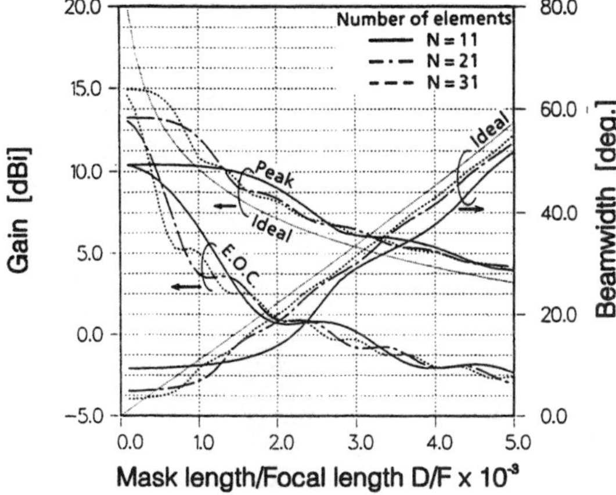

Figure 7.10 Plots of gain and beamwidth versus *D/F* for N = 11, 21, and 31. (© 1992 IEEE.)

7.10, "peak" and "E.O.C." denote the peak gain and the gain at the edge coverage, respectively.

Plots of gain and beamwidth versus *D/F* for the Gaussian P_2 distribution of the *edge level* (EL) 0 dB, −5 dB, −10 dB, and −15 dB are shown in Figure 7.11.

The low sidelobe level of the array can be achieved by reducing the edge level of the array aperture amplitude distribution, which is controlled by the reference beam P_2. The amplitude distribution for the reference beam P_2 is in Gaussian form, and Figures 7.12 and 7.13 show plots of the sidelobe level versus *D/F* as functions of edge level for zero offset and 250-μm offset (P_2-axis), respectively. For these plots, the number of antenna

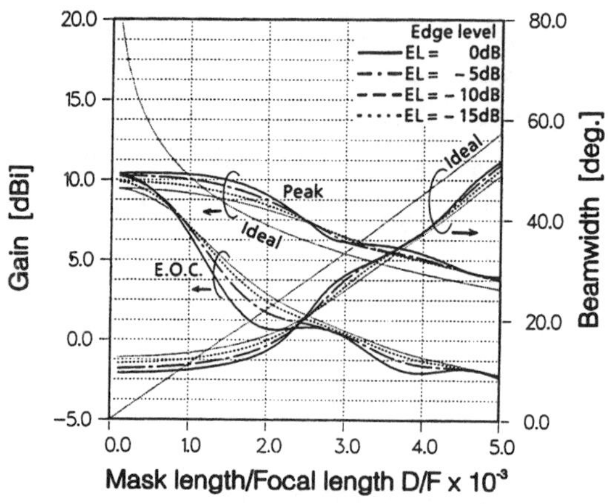

Figure 7.11 Gain and beamwidth versus *D/F* for the Gaussian reference beam. (© 1992 IEEE.)

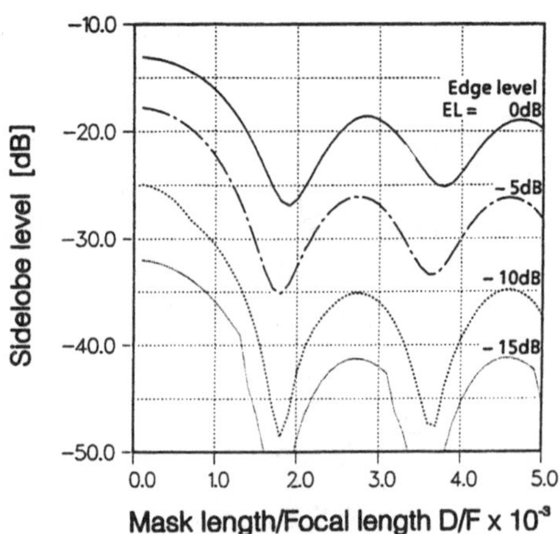

Figure 7.12 Sidelobe level versus *D/F* for the Gaussian reference beam. (© 1992 IEEE.)

elements N is 11. It is shown that the sidelobe level decreases with the edge level. Due to the asymmetrical distribution in the case of offset on the P_2-axis by 250 μm, the sidelobe level for EL = −15 dB becomes about 7 dB higher than that shown in Figure 7.12. Figure 7.14 shows plots of sidelobe level versus *D/F* for a Gaussian error in aperture amplitude (ΔA) and phase ($\Delta \phi$). In the figure, curves (a) represent the amplitude error of 1.5 dBrms and phase error of 2.5 degrms and curves (b) represent amplitude error of 3 dBrms and phase error of 5 degrms as functions of EL.

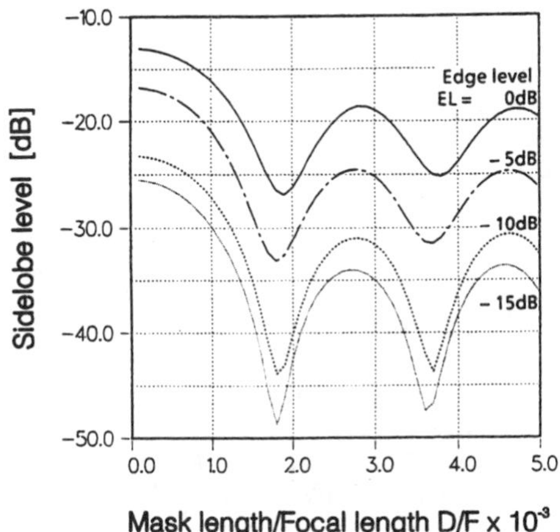

Figure 7.13 Plots of sidelobe level versus *D/F* for the offset reference beam. (© 1992 IEEE.)

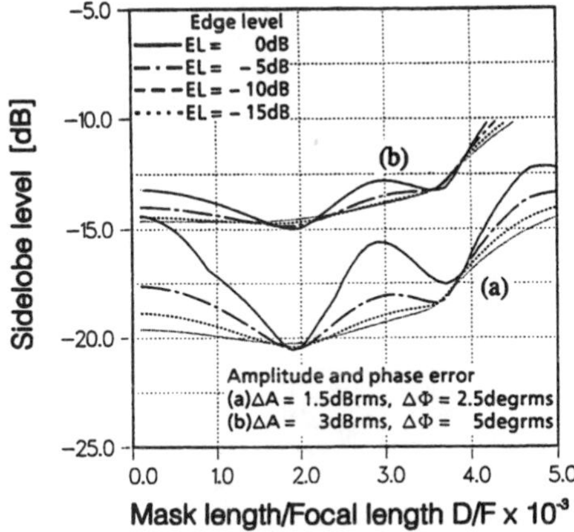

Figure 7.14 Plots of the sidelobe level versus *D/F* for the Gaussian error. (© 1992 IEEE.)

References

[1] Konishi, Y., W. Chujo, H. Iwasaki, and K. Yasukawa, ''A Study of Radiation Characteristics in Optically Controlled Array Antenna,'' *Proc. Int. Symp. Antenna and Propagation (ISAP),* Japan, 1989, pp. 965–968.

[2] Konishi, Y., W. Chujo, H. Iwasaki, and K. Yasukawa, ''Techniques for Antenna Sidelobe Suppression and Laser Power Reduction in Optically Controlled Array Antenna,'' *Proc. SPIE Optoelectronic Signal Processing for Phased Array Antennas II,* Vol. 1217, 1990, pp. 26–34.

[3] Konishi, Y., W. Chujo, and M. Fujise, ''Beam Scanning Characteristics in an Array Antenna with BFN Using Fourier Optics and Photomixing,'' *Proc. SPIE,* Vol. 1703, 1992, pp. 438–447.

[4] Konishi, Y., W. Chujo, and M. Fujise, ''Carrier-to-Noise Ratio and Sidelobe Level in a Two-Laser Model Optically Controlled Array Antenna Using Fourier Optics,'' *IEEE Trans. Ant. Prop.,* Vol. AP-40, No. 12, December 1992, pp. 1459–1465.

[5] Pan, J. J., ''Fiber Optics for Microwave/Millimeter-wave Phased Array,'' *Microwave System News and Communication Technology,* July 1989, pp. 48–54.

[6] Pan, J. J., and W. Z. Li, ''High On-Off Ratio, Ultrafast Optical Switch for Optically Controlled Phased Array,'' *Proc. SPIE,* Vol. 1476, 1991, pp. 122–131.

[7] Pan J. J., S. L. Chia, and W. Z. Li, ''Cost-Effective Optical Switch Matrix for Microwave Phased Array,'' *Proc. SPIE,* Vol. 1476, 1991, pp. 133–142.

[8] Koepf, G. A., ''Optical Processor for Phased Array Antenna Beam Formation,'' *Proc. SPIE,* Vol. 477, May 1984, pp. 75–81.

[9] Anderson, L. P., F. Boldisser, and D. C. D. Chang, ''Antenna Beamforming Using Optical Processing,'' *Proc. SPIE,* Vol. 886, January 1988, pp. 228–232.

[10] Nickerson, K. A., P. E. Jessop, and S. Haykin, ''Optical Processor for Array Antenna Beam Shaping and Steering,'' *Proc. SPIE,* Vol. 1217, January 1990, pp. 184–195.

[11] Yamada, K., Y., W. Chujo, I. Chiba, M. Fujise, and Y. Konishi, ''Experimental Study on C/No in an Optically-Controlled Array Antenna,'' *Proc. Asia-Pacific Microwave Conference,* Vol. 2, National Chiao Tung University, Hsinchu, Taiwan, October 1993, pp. 18-12–18-15.

Appendix
History of Optical Communication Technology

Year	Description of Developments
1850s	Light guidance demonstration.
1870s	Alexander Graham Bell's early experiments with optical communications in the form of "photo phone."
1910	Electromagnetic theory of microwave dielectric rods and optical fiber systems.
1930–1950	Fabrication and experimentation on the bundles of glass filaments.
1960	Publication of the theory of fiber modes.
1960s	Light was guided experimentally by lens trains, hollow tubes, and fibers.
1962	Fabrication experiments began on semiconductor injection laser diodes.
1966	Dielectric-fiber surface waveguides for optical frequencies (single-mode fibers).
1968	Fabrication and experiments began on the graded-index multimode fibers that provided a loss of about 1 dB/m.
1970	Manufacturing of single-mode fibers at a wavelength of 633 μm for a distance of 20 km. Fabrication and experimentation on several 850-nm components.
1971	Prediction of infinite chromatic bandwidth near 1.3 μm.
1972	Single-mode polarization properties.
1973–1975	Multimode fiber, prototype components, several subkilometer links demonstration. Work began on fiber optic sensors.
1976	Demonstration at a wavelength of 1.3 μm for a loss of 0.5 dB/km. Telephony trials at 0.85 μm with repeater spacing of a few kilometers. Work started on directional couplers and wavelength-division multiplexing. Research began on phased array scanning and sequential scanning of multielement antenna systems, and the ideas to introduce fibers in beamforming networks were produced.
1977–1978	Trial on interoffice trunking, subscriber loops, and cable TV using multimode fiber. Material research for 2-μm to 12-μm wavelengths to provide attenuation below 0.01 dB/km.
1979	Many countries (Asia, Australia, Europe, and North America) install multimode fiber links at a wavelength of 850 nm. Demonstration on the attenuation of fibers (0.2 dB/km) at 1.55 μm. Coherent transmission through fibers started.
1980	Trials started with InGaAsP and Ge APDs at a wavelength of 1.3 μm.
1981–1982	Megameter quantities of a multimode fiber were used for links up to 90 Mb/s over 30 km. Optical fiber was used to cover broadband networks, submarine links, and local area networks. Single-mode laboratory demonstrations for more than 100-km spacing at several hundred Megabytes per second.
1983	United Kingdom and United States operated the first commercially operating single-mode optical fibers (SMF) links, and very large manufacturing started on SMF.
1984	Optical fibers for telephone lines began to be placed in many countries all over the world. Research began on beamforming networks in many countries.
1985–1986	Ultra-high-speed semiconductor lasers. Optical tuning of MESFET. Optical switching of devices.
1987–1994	High-speed modulation of semiconductor lasers. Beamforming networks for radar and satellite antennas. Low-loss optical fibers. Integration of lasers and microwave and millimeter-wave semiconductor devices for high-speed systems and data processing systems. Semiconductor optical fiber amplifiers. Quantum wire and dot lasers. Optical parametric oscillators. 5-Gb/s optically amplified transoceanic undersea systems. Work on 10-Gb/s to 100-Gb/s transmission systems. Figures 1.1 and 1.3 show developments in sources and achievements on optical transmission capacity.

About the Author

Akhileshwar Kumar earned his Ph.D. in Engineering at the University of Strathclyde in Glasgow, United Kingdom. He currently serves as president of AK Electromagnetique, Inc., Quebec, Canada, and has designed various satellite antennas including a telemetry antenna for ESA's ERS-1 and a mobile terminal antenna for the MSAT and INMARSAT applications. He has also developed an optical fiber beamforming network for satellite antennas. He has published three additional books: *Radar Absorbing Materials* (1982), *Microwave Cavity Antennas* (Artech House, 1989), and *Fixed and Mobile Terminal Antennas* (Artech House, 1991). He has published more than 115 technical papers and is a senior member of the IEEE.

Index

The Artech House Antenna Library

Helmut E. Schrank, *Series Editor*

Advanced Technology in Satellite Communication Antennas: Electrical and Mechanical Design, Takashi Kitsuregawa

Analysis Methods for Electromagnetic Wave Problems, Volume 2, Eikichi Yamashita, editor

Analysis of Wire Antennas and Scatterers: Software and User's Manual, A. R. Djordjević, M. B. Bazdar, G. M. Bazdar, G. M. Vitosevic, T. K. Sarkar, and R. F. Harrington

Analysis Methods for Electromagnetic Wave Problems, E. Yamashita, editor

Antenna-Based Signal Processing Techniques for Radar Systems, Alfonso Farina

Antenna Design With Fiber Optics, A. Kumar

Broadband Patch Antennas, Jean-François Zürcher and Fred E. Gardiol

CAD for Linear and Planar Antenna Arrays of Various Radiating Elements: Software and User's Manual, Miodrag Mikavica and Aleksandar Nešić

The CG-FFT Method: Application of Signal Processing Techniques to Electromagnetics, Manuel F. Cátedra, Rafael P. Torres, José Basterrechea, Emilio Gago

Electromagnetic Waves in Chiral and Bi-Isotropic Media, I.V. Lindell, S.A. Tretyakov, A.H. Sihvola, A. J. Viitanen

Fixed and Mobile Terminal Antennas, A. Kumar

Generalized Multipole Technique for Computational Electromagnetics, Cristian Hafner

Handbook of Antennas for EMC, Thereza Macnamara

Integral Equation Methods for Electromagnetics, N. Morita, N. Kumagai, and J. Mautz

IONOPROP: Ionospheric Propagation Assessment Program, Version 1.1: Software and User's Manual, by Hernert V. Hitney

Four-Armed Spiral Antennas, Robert G. Corzine and Joseph A. Mosko

Introduction to Electromagnetic Wave Propagation, Paul Rohan

Introduction to the Uniform Geometrical Theory of Diffraction, D. A. McNamara

Microwave Cavity Antennas, A. Kumar and H. D. Hristov

Millimeter-Wave Microstrip and Printed Circuit Antennas, Prakash Bhartia

Mobile Antenna Systems, K. Fujimoto and J. R. James

Modern Methods of Reflector Antenna Analysis and Design, Craig Scott

Moment Methods in Antennas and Scattering, Robert C. Hansen, editor

Monopole Elements on Circular Ground Planes, M. M. Weiner et al.

Near-Field Antenna Measurements, D. Slater

Passive Optical Components for Optical Fiber Transmission, Norio Kashima

Phased Array Antenna Handbook, Robert J. Mailloux

Polariztion in Electromagnetic Systems, Warren Stutzman

Practical Phased-Array Antenna Systems, Eli Brookner et al.

*Practical Simulation of Radar Antennas and Radome*s, Herbert L. Hirsch and Douglas C. Grove

Radiowave Propagation and Antennas for Personal Communications, Kazimierz Siwiak

Reflector and Lens Antennas: Software User's Manual and Example Book, Carlyle J. Sletten, editor

Reflector and Lens Antenna Analysis and Design Using Personal Computers, Carlyle J. Sletten, editor

Shipboard Antennas, Second Edition, Preston Law

Small-Aperture Radio Direction-Finding, Herndon Jenkins

Spectral Domain Method in Electromagnetics, Craig Scott

Understanding Electromagnetic Scattering Using the Moment Method: A Practical Approach, Randy Bancroft

Waveguide Components for Antenna Feed Systems: Theory and CAD, J. Uher, J. Bornemann, and Uwe Rosenberg

For further information on these and other Artech House titles, contact:

Artech House
685 Canton Street
Norwood, MA 02062
617-769-9750
Fax: 617-769-6334
Telex: 951-659
e-mail: artech@world.std.com

Artech House
Portland House - Stag Place
London SW1E 5XA England
+44 (0) 171-973-8077
Fax: +44 (0) 171-630-0166
Telex: 951-659
e-mail: bookco@artech.demon.co.uk